Deconstructing Complexity: Biophysical Characterization of Protein-Lipid Membranes

Parkar

CHAPTER 1

INTRODUCTION

1.1. Characterization of biological systems

Biological systems are regulated by an incredibly complex, fine-tuned set of systems that has been evolutionarily optimized, and the treatment of illness requires a detailed understanding of the connections and mechanisms behind the moving pieces to understand their dysfunction. Many approaches can be (and should be) taken to understand this detail as each approach carries its own benefits, where some approaches are better suited for global readout, such as a patient's health, while others are optimized to report on a specific detail, such as protein structure. Regardless of the method, interpretation of an experiment requires a thorough understanding of how the experiment's readout relates the studied factors to the underlying system.

We approach our investigations into protein:lipid membrane systems through a biophysical characterization of the direct components involved in the interaction. In doing so, we sacrifice the complexity of the full set of variables of the biological system that undoubtedly influence these interactions. But through our methods we are able to uncover specific, molecular details of the underlying systems that can, in turn, be integrated with this more complete parameter set. Our approach additionally emphasizes how intricately tied biological systems and the mathematical frameworks used for their characterization are. Ultimately, the work in this book aims to demonstrate the utility of biophysical characterization to further our understanding of physiological processes and corresponding disease states.

1.2. Organization

The following work in this book is organized by the specific aims of the two protein systems studied in this book: α-synuclein and the human Transmembrane Immunoglobulin and Mucin

Domain-Containing Protein-3 (hTIM3). **Chapter 2** discusses the biological background and the characterization framework of these two proteins, which is followed by the methodology and protocols of the techniques that were used to study these protein systems in **Chapter 3**. Our primary results of the structural detail of the α-synuclein:membrane system related to disease state are described in **Chapter 4**, while **Chapter 5** contains a discussion of preliminary data that points to the protein concentration-dependent binding of α-synuclein. The final chapter of this book, **Chapter 6**, summarizes our work into the structural characterization of the membrane-binding of hTIM3.

CHAPTER 2

**BACKGROUND: LIPID:PROTEIN CONTACTS IN THE BINDING OF α-SYNUCLEIN
AND hTIM3 TO LIPID MEMBRANES**

A complex set of cellular processes regulates protein:membrane systems to allow for intercellular communication and function. Within this, there are many factors and features of the systems that are carefully controlled and modified in response to both internal and external signaling. This chapter discusses the lipid membrane parameters that are relevant to the protein systems studied within this book. It additionally covers the functional role and known structural features of α-synuclein (**Chapters 4** and **5**) and the transmembrane immunoglobulin and mucin domain-containing protein-3 (TIM3) (**Chapter 6**) protein systems with an emphasis on their lipid membrane-binding role.

2.1. The Lipid Membrane and Peripheral Membrane Proteins

Cells and organelles separate their contents from the external aqueous environment through a membrane whose basic structure is a lipid bilayer. Lipids are amphipathic molecules, composed of a hydrophobic tail and a hydrophilic head group, resulting in the self-assembly in an aqueous environment of a bilayer structure where the hydrophobic core acts as barrier between the external and internal contents of cells and organelles[1] (**Fig. 2.1**). While this lipid bilayer is often depicted as a uniform layer of identical lipids, the true membrane is composed of a variety of lipids and proteins where the cell expends energy to maintain specific compositions on both the inner and outer leaflets of the membrane.[2] These specific lipid compositions can inform on a cell's status or health.[3,4]

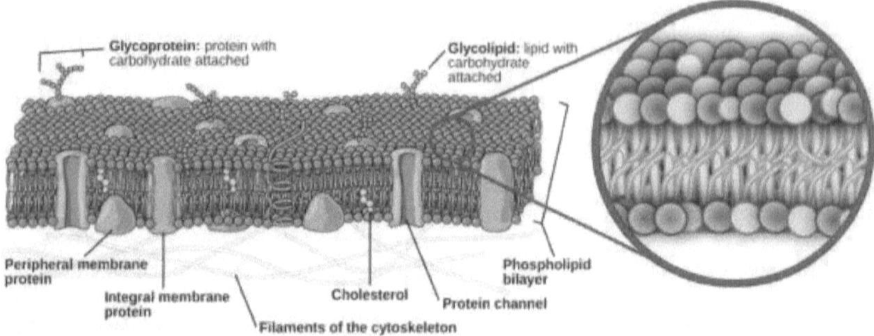

Fig. 2.1. Depiction of lipid membrane components within the lipid bilayer. Zoom-in of lipid membrane shows the variety of lipid headgroups and lipid tails of the membrane. Left portion of figure adapted from Molnar *et al.*[1]

Differences in the tails and headgroups of lipids allow for local control over the physical properties of the membrane. For example, lipid tails can have varying lengths and degrees of saturation, which can alter membrane fluidity and the presence of membrane defects. Meanwhile, differences in the charge and size of the lipid headgroups alter the lipid landscape that is exposed to the external aqueous environment (**Figs. 2.1, 2.2**). Lipid membranes can additionally hold up to ~30% molar cholesterol.[5] While its sterol core differs from the general structure of lipid molecules (**Fig. 2.2**), cholesterol (Chol) plays a large role in modulating the packing and fluidity of the membrane.[6] The cell's control over the complexity of these lipid compositions plays a mechanistic role within inter- and intra-cellular signaling, where protein recognition of specific membranes allows for selectivity in initiating downstream processes.

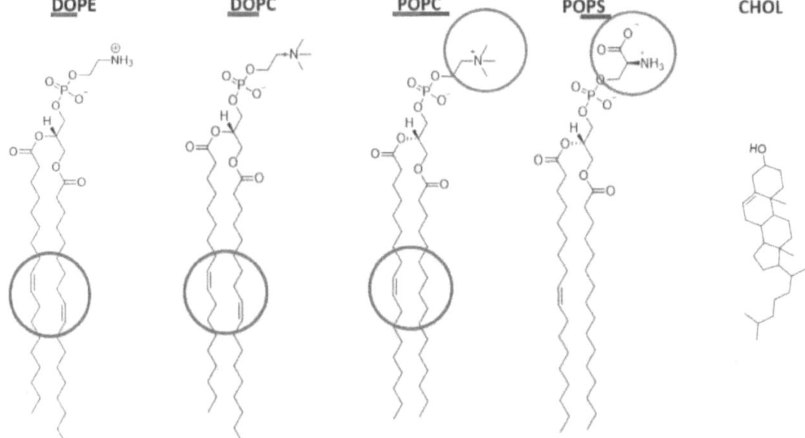

Fig. 2.2. Examples of differences in tails and headgroups of lipid molecules and the structure of cholesterol (Chol). Differences in tail saturation are shown in red between dioleoyl (DO) and palmitoyl-oleoyl (PO) lipids. Differences in head size and charge are shown in blue between phosphatidylcholine (PC) and phosphatidylserine (PS), and phosphatidylethanolamine (PE).

Peripheral membrane proteins are a type of membrane-binding proteins that play a role within cellular signaling. Generally, this subset of proteins only partially and temporarily associates with the lipid membrane (**Fig. 2.3**), stimulating downstream pathways through a variety of mechanisms including conformational changes, lipid rearrangements, and formation of protein complexes.[7] The lipid-bound states of these proteins can provide key insights into cellular signaling pathways and associated disease states. While the bound states of peripheral membrane proteins are often focused on the contacts between the protein and a single lipid molecule, the actual interactions often span across a much larger lipid surface area, where additional protein:lipid contacts affect protein structure and orientation. In turn, these two traits of the bound state dictate protein function, dysfunction, and signal transduction based on contextual membrane binding. Characterizing these system parameters is critical in both understanding a protein's role within cellular processes and aiding the development of drugs and antibodies. Two such protein systems are α-synuclein and TIM3, where protein:membrane interactions are proposed to play a central role within immune exhaustion[8] and Parkinson's Disease,[9] respectively.

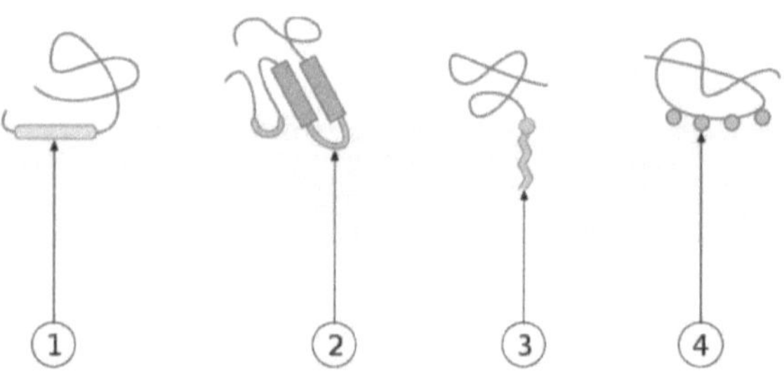

Fig. 2.3. Examples of the various states of insertion and association of peripheral membrane-binding proteins.[10] Membrane-binding of α-synuclein and TIM3 are most similar to Examples 1 and 2, respectively, where α-synuclein forms two α-helices that partially inserts into the membrane and hTIM3 inserts its loop regions.

2.2. α-synuclein

2.2.1. Biological Role of α-synuclein

α-synuclein is a lipid-binding protein that plays a role in Parkinson's Disease (PD), a neurodegenerative disease, primarily characterized by a decline in motor and cognitive function.[11] Though not deadly, PD progression results in severe impacts on quality of life with no existing cures or preventative treatments. The disease affects over half a million adults in the United States with onset at an average age of 62 years.[12] However, early-onset PD has been observed in patients with genetic markers within the SNCA gene.[13]

The SNCA gene codes for the protein α-synuclein, which has been directly linked to irregular protein aggregation and formation of toxic Lewy bodies (LBs) within neurons – a common marker of PD.[9] Though α-synuclein's function within the neuron is not fully understood, it is known to bind to the lipid membrane of synaptic vesicles (SVs), where it is proposed to play a role in the trafficking of SVs and neurotransmitter release into the synaptic cleft[14] (**Fig. 2.4a**). SVs are small intraneural organelles that, similarly to cells, have a lipid bilayer that differentiates their identity from other lipid membranes of or in the neuron. Intraneuronal communication is a highly regulated process, necessitating control over both the temporal and spatial release of neurotransmitters, and one can imagine that disruption to the process would result in an impairment of normal body or cognitive function.[15]

Dysregulation in protein:membrane binding can lead to lipid membrane-templated α-synuclein fibrillization,[16] and has been proposed to result in LB formation[17,18] (**Fig. 2.4b**), where the primary components of LBs are α-synuclein and other protein and lipid species found within the neuron and the membrane of SVs.[17,19] These aggregate species and LBs likely contribute to the decline in cognitive and motor function of patients with PD, and therefore uncovering the molecular basis and structural detail of α-synuclein's role within this state may provide insight into the prevention or treatment of the disease.

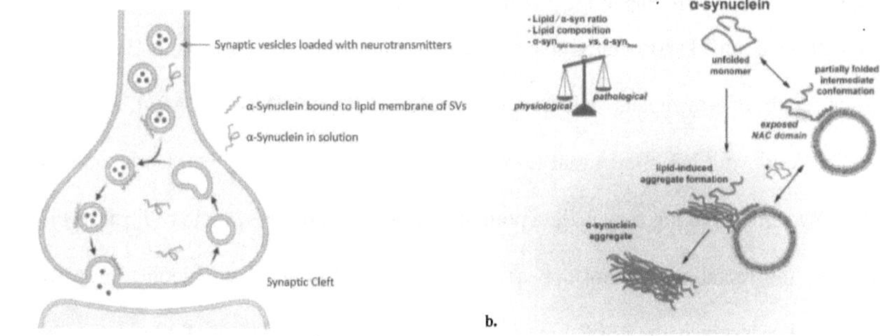

Fig. 2.4. Physiological role of α-synuclein and disease associated aggregation.

In the cytosol of the neuron, α-synuclein has little secondary structure, forming an intrinsically disordered protein (IDP).[21] However, in the presence of lipid membranes, its N-terminus (residues 1-95) forms an extended α-helical region that partially inserts into the membrane,[22] while its C-terminus (residues 96-140) remains disordered and forms the projecting domain. This lipid-binding region of α-synuclein can be treated as two α-helices, where Helix 1 roughly comprises residues ~1-37 and Helix 2 of residues ~45-95[23,24] (**Fig. 2.5**).

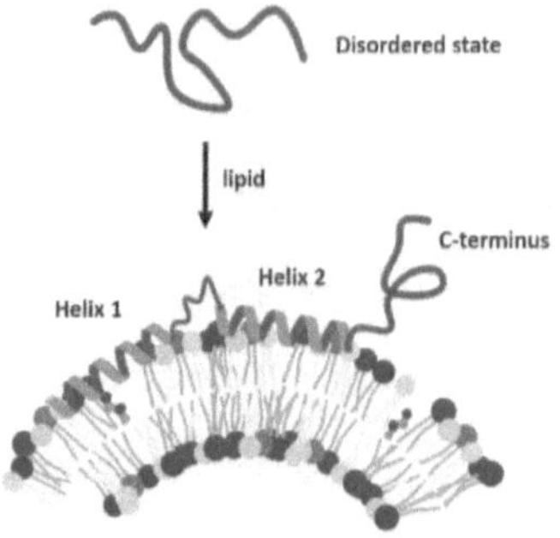

Fig. 2.5. Transition of α-synuclein from its disordered solution state to the partial α-helical lipid-bound state. The C-terminus of the protein projects off the membrane in the lipid-bound state.

Of importance, all PD-associated genetic point mutations in SNCA are found within the lipid-binding region of α-synuclein: V15A, A30P, E46K, H50Q, G51D, A53E, A53T, and A53V[13](**Fig. 2.6**). Previous work has shown that fibrillization of α-synuclein may be membrane-templated,[25] with a few studies linking α-synuclein point mutations to an increased rate of fibril formation.[26,27] However, the mutations do not appear to uniformly affect the lipid-binding properties of α-synuclein. For instance, the A30P mutant decreases α-synuclein's affinity for lipid membranes,[27,28] while the E46K increases its affinity.[29] Additionally, there are only limited data sets on how the mutations differentially affect each helix of the protein. As binding studies of α-synuclein are sensitive to experimental conditions, it is difficult to compare the effects of the mutations across studies. There is a need for a comprehensive, uniform analysis of how all the PD-associated point mutations affect interactions between α-synuclein and lipid membranes with helix-specific resolution to uncover the roles of the individual helices in the progression of PD.

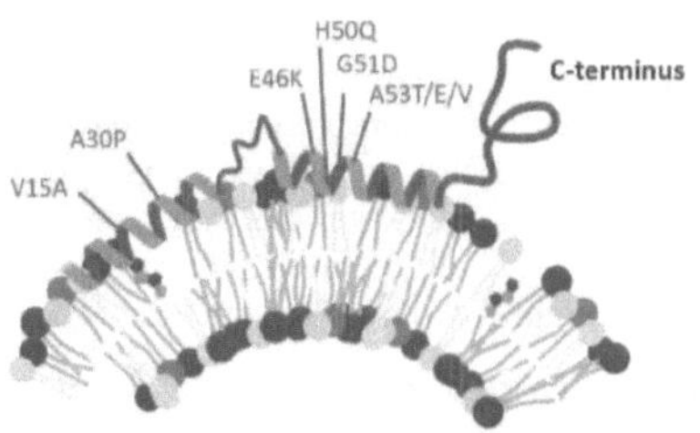

Fig. 2.6. PD-associated mutations in α-synuclein are all located within the lipid-binding region.

2.2.2. Lipid-Bound State of α-synuclein

While the two helices of α-synuclein are separated by a region that remains disordered, the binding of each helix is still likely dependent on the bound state of the other. Both helices are amphipathic,[22] composed of imperfect repeats of the KTKEGV motif,[30] allowing residues to form hydrophobic contacts with lipid tails and hydrophilic contacts with the aqueous environment outside the synaptic vesicle (**Fig. 2.7a,b**). Of note, the boundary of the hydrophilic and hydrophobic regions of the vesicles are lined with lysine residues (shown in blue in **Fig. 2.7**). These positively charged residues can interact with negatively charged lipid headgroups, such as those of phosphatidylserine (PS), further stabilizing the helices on the surface of the membrane.[22]

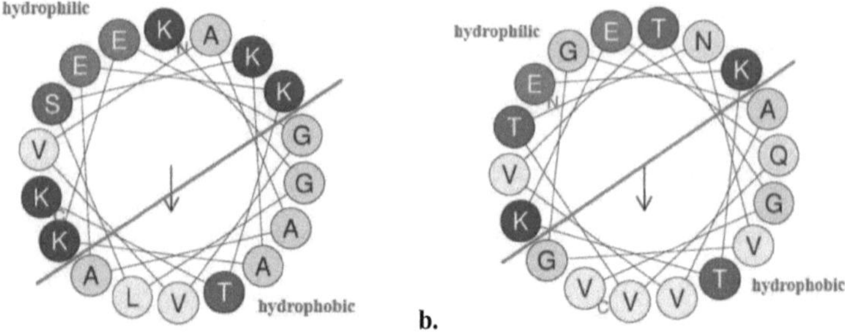

Fig. 2.7. Amphipathic properties of sample regions of the two α-helices of α-synuclein **a.** Residues 6-23 of Helix 1. **b.** Residues 57-74 of Helix 2. Positively and negatively charged residues are shown in blue and red, respectively. Nonpolar residues are shown in yellow and gray. Polar residues are shown in pink and purple. Red line delineates between hydrophobic and the hydrophilic regions of the protein. Figures made using heliQuest α-helix prediction tool.

The insertion of the helices of α-synuclein requires rearrangement of the lipid membrane to accommodate the large surface area of the protein. Despite being a relatively small protein (~15 kDa), a full extension of its helices would cover ~14 nm.[31] The disordered region between the helices has been shown to allow the protein to adopt various states (e.g., bent or antiparallel helices),[31] but nonetheless the lipid-bound state of the protein would involve many lipid contacts along the helices. Therefore, while individual lipid species, such as PS, have been shown to increase the affinity of α-synuclein for the membrane,[32] the increase cannot be isolated to the effect of any one individual protein:lipid contact, and is likely the product of multiple different contact sites as well as overall changes to the membrane surface.

SVs are smaller organelles with a diameter of ~40 nm,[33] resulting in a highly curved surface. Additionally, their membranes are composed of polar and negatively charged headgroups, primarily including PE, PC, PS, and Chol with smaller amounts of phosphatidylinositol and sphingomyelin.[34] Lipids of SV membranes are enriched in poly-unsaturated tails,[35] likely increasing membrane fluidity and the space between lipid molecules. The binding of α-synuclein has been previously shown to be dependent on these membrane parameters, favoring large membrane curvature (with increased binding on smaller vesicles),[36] negative charge,[32] and greater lipid tail unsaturation.[37] Overall, these features indicate that the binding of α-synuclein is largely dependent on the ability of a lipid membrane to accommodate both the insertion of the hydrophobic portion of α-synuclein and the stabilization of the protein through polar contacts.

As polyunsaturated tails are susceptible to oxidation, DO lipid tails can be used in their place in model systems to mimic the unsaturation of lipid tails while retaining membrane stability. Within the work in this book, we mimic the lipid membrane of SVs through a 55:20:15:10 DOPC:DOPS:DOPE:CHOL model.

The characterization of α-synuclein has been done through a variety of methods[38] and has largely focused on the structural properties of its monomer and oligomer forms.[39] While a few of these methods, such as cryo-electron microscopy,[4] can probe at the structures of the lipid-bound states, the wide set of parameters affecting the binding of α-synuclein complicates our understanding of its lipid-binding role within disease progression and the factors underlying PD-related protein aggregation. In addition to exploring the effect of PD-associated mutations on the lipid-bound structure of α-synuclein as described above, the work within this book aims to discuss the ability of various lipid-binding assays to quantify α-synuclein:lipid binding in congruence with structural detail.

2.3. Transmembrane Immunoglobulin and Mucin Domain-Containing Protein-3

2.3.1. Biological Role of hTIM3

The TIMs are a family of PS-recognizing proteins that is composed of four murine (mTIM1, mTIM2, mTIM3, and mTIM4) and three human (hTIM1, hTIM3, and hTIM4) members.[41] These proteins are expressed on a variety of immune cells with a range of proposed functions within various illnesses, including cancer[42,43] and asthma,[44] along with clearance of apoptotic cells.[45] In fulfilling their role within the immune system, one of the primary functions of the TIMs is the recognition of target membranes. As membranes of target cells can undergo lipid compositional changes corresponding to their state,[46] selective recognition and binding of the protein to lipid membranes can result in state-specific downstream responses. For example, cells undergoing apoptosis will intentionally scramble their membranes through the activation of scramblases and deactivation of flippases,[47] exposing PS and other negatively charged lipids on the membrane's outer leaflet. Similarly, cancerous cells can lose their ability to regulate lipid asymmetry,[48] resulting in PS-exposure on the outer leaflet (**Fig. 2.8**). The exposure of PS can then be recognized by TIM3 expressed on the cell surface of macrophages, which in turn stimulates the specific engulfment and disposal of apoptotic cells, allowing for differentiation from healthy cells.[45]

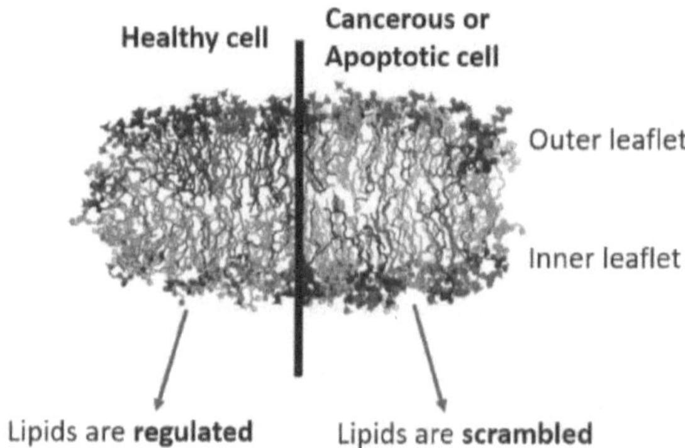

Fig. 2.8. Depiction of scrambling of the lipid membrane that results in a loss of membrane asymmetry and the exposure of PS (green) on the outer leaflet of the membrane in cancer or during apoptosis. Figure adapted from Rivel *et al.*[46]

TIM3 has been linked to both stimulatory and inhibitory functions within T-cell activity in the immune response, where its function may be dependent on co-stimulatory or co-inhibitory binding partners. Two such direct binding partners are the lipid PS and the protein galectin-9.[8] Additionally, the indirect blocking of other co-inhibitory proteins, such as programmed cell death protein-1[49] (PD-1), can upregulate the expression of hTIM3. While the full pathways are not understood, prolonged stimulation of regulated T-cells by TIM3 is linked to apoptosis of T-cells and immune exhaustion within a wide range of cancers and chronic viral infections, including liver,[43] breast,[50] and brain cancer.[51] This suggests that TIM3 plays a regulatory role within the progression of these illnesses and may be an important drug target in safeguarding the body's natural immune response. Over the past few years, clinical research has taken an increased interest in the antibody targeting of TIM3, and several preliminary studies have found that the targeting of the protein improves patient outcomes by preventing T-cell exhaustion. Importantly, studies have implicated the binding of TIM3 to PS within immune cell regulation,[52] and structural studies of successful antibodies that recovered immune exhaustion found epitopes that overlapped with the PS-binding site of TIM3[53–55] (**Fig. 2.9**). While the mechanism and the cellular process behind these antibodies are unclear, these details highlight the importance of characterizing the structural features of TIM3 involved in PS-binding, as it may allow for increased selectivity and efficacy in drug development.

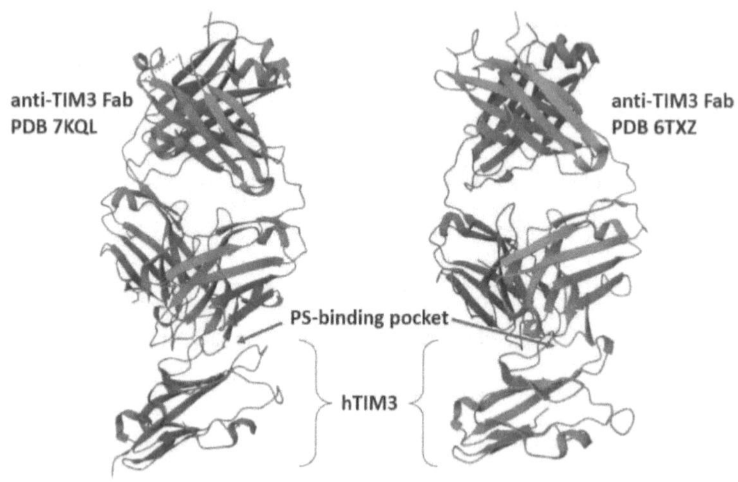

Fig. 2.9. Published structures of anti-hTIM3 antibodies, **PDB 7KQL**[54] (left) and **PDB 6TXZ** (right)[55], showing the Fab's recognition of the PS-binding pocket of hTIM3.

2.3.2. Lipid:Protein Contacts in the PS-Binding of hTIM3

The structures of the TIMs include a cytoplasmic tail involved in downstream signaling, a transmembrane domain, and an extracellular domain that is composed of a mucin stalk and an immunoglobulin variable (IgV) domain (**Fig. 2.10**).[8] These IgV domains, with the exception of mTIM2, contain a PS-binding pocket that coordinates a calcium ion, which forms contacts with residues within the pocket and the negatively charged headgroup of PS.[8]

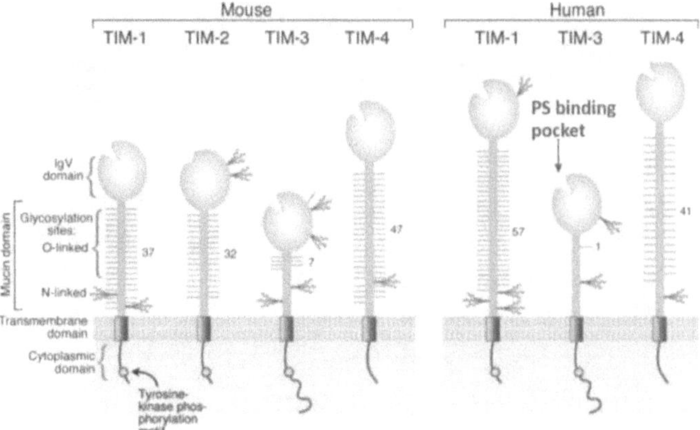

Fig. 2.10. Structural features of the TIM family of protein. The PS-binding pocket of hTIM3 is indicated by the red arrow. Figure adapted from Freeman *et al.*[41]

While this binding pocket accommodates a single PS lipid molecule, the true binding of the IgV domain to PS can be more accurately characterized as an ensemble of bound states that involves additional lipid contacts occurring between the domain and the surface area that the protein occupies on the lipid membrane[56] (**Fig. 2.11**). Among others, this includes hydrophobic contacts of hydrophobic residues with lipid tails and polar contacts between peripheral residues with lipid headgroups.[56] These protein:lipid contacts contribute to the stability and fluctuations of the protein at the membrane, affecting binding features such as protein orientation at the membrane. By exploring the binding of the IgV domains through these additional contacts, we can more thoroughly characterize the structural features of the bound state of the protein that may be essential to understanding its PS-binding mechanism and subsequent signal transduction.

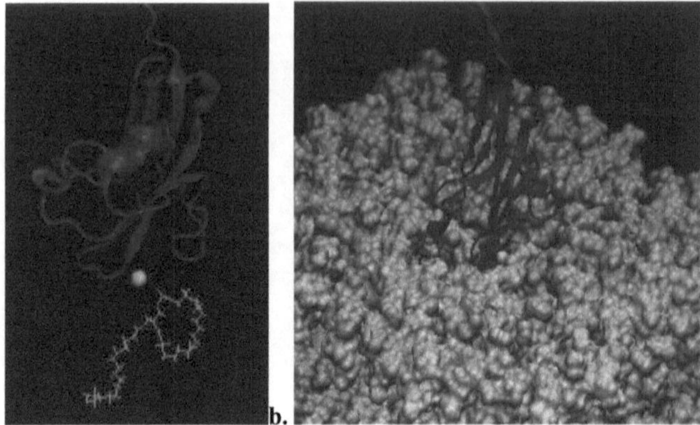

Fig. 2.11. Depiction of lipid:protein contacts in hTIM3:membrane binding. **a.** A single PS forms contacts with the Ca^{2+} ion and residues of the binding pocket of hTIM3. **b.** Membrane-inserted state of hTIM3, showing additional lipid contacts that form. Cartoon representation of protein is shown in blue, Ca^{2+} ion is shown in yellow, lipids are shown in surface representation with PS in pink and PC in gray. Membrane in panel **b** is rotated ~45° from top-down view for better visualization of protein:lipid contacts.

Previous work by our group had characterized the murine proteins' affinities for Ca^{2+} and PS, and identified peripheral residue contacts that contribute to protein stabilities on the membrane surface.[56,57] We have since then extended the developed methods to study the human TIM3 variant (hTIM3) due to the field's increased interest in it as a therapeutic target within cancer and other illnesses. While hTIM3's other binding partners, such as PD1 – a checkpoint protein on T cells, are often included in functional assays, PS is an understudied ligand despite functional importance. Several crystal structures of hTIM3 have been published, including ones in complex with small molecules or antibodies, but its membrane-bound state has not yet been identified. This state could be critical to therapeutic development, as recent molecular dynamics (MD) simulations suggest that the protein may undergo a conformational switch upon engagement of its pocket with a short-tailed PS molecule[58] (the structural features of this conformational switch are more thoroughly discussed in **Chapter 6**). Our work here aims to 1) characterize the protein:lipid contacts in a membrane-bound state of hTIM3, and 2) determine if this conformational switch occurs for the membrane-bound state of hTIM3. Through this, we aim to provide the structural basis for increasing the efficacy of existing drug treatments through the selective targeting of a solution or a membrane-bound state.

2.4. References

1.	Molnar, C. & Gair, J. *Concepts of Biology*. (BC Open Textbook Project, 2015).

2.	Hankins, H. M., Baldridge, R. D., Xu, P. & Graham, T. R. Role of Flippases, Scramblases and Transfer Proteins in Phosphatidylserine Subcellular Distribution. *Traffic* **16**, 35–47 (2015).

3.	Magtanong, L., Ko, P. J. & Dixon, S. J. Emerging roles for lipids in non-apoptotic cell death. *Cell Death Differ* **23**, 1099–1109 (2016).

4.	Szlasa, W., Zendran, I., Zalesińska, A., Tarek, M. & Kulbacka, J. Lipid composition of the cancer cell membrane. *J Bioenerg Biomembr* **52**, 321–342 (2020).

5.	Zhang, J. *et al.* Cholesterol content in cell membrane maintains surface levels of ErbB2 and confers a therapeutic vulnerability in ErbB2-positive breast cancer. *Cell Communication and Signaling* **17**, 15 (2019).

6.	Chakraborty, S. *et al.* How cholesterol stiffens unsaturated lipid membranes. *Proceedings of the National Academy of Sciences* **117**, 21896–21905 (2020).

7.	Boes, D. M., Godoy-Hernandez, A. & McMillan, D. G. G. Peripheral Membrane Proteins: Promising Therapeutic Targets across Domains of Life. *Membranes (Basel)* **11**, 346 (2021).

8.	Kane, L. P. T Cell Ig and Mucin Domain Proteins and Immunity. *The Journal of Immunology* **184**, 2743–2749 (2010).

9.	Siddiqui, I. J., Pervaiz, N. & Abbasi, A. A. The Parkinson Disease gene SNCA: Evolutionary and structural insights with pathological implication. *Sci Rep* **6**, 24475 (2016).

10.	Peripheral membrane protein. *Wikipedia* https://en.wikipedia.org/wiki/Peripheral_membrane_protein (2023).

11.	Armstrong, M. J. & Okun, M. S. Diagnosis and Treatment of Parkinson Disease. *JAMA* **323**, 548 (2020).

12.	Pagano, G., Ferrara, N., Brooks, D. J. & Pavese, N. Age at onset and Parkinson disease phenotype. *Neurology* **86**, 1400–1407 (2016).

13. Whittaker, H. T., Qui, Y., Bettencourt, C. & Houlden, H. Multiple system atrophy: genetic risks and alpha-synuclein mutations. *F1000Res* **6**, 2072 (2017).

14. Gao, V., Briano, J. A., Komer, L. E. & Burré, J. Functional and Pathological Effects of α-Synuclein on Synaptic SNARE Complexes. *J Mol Biol* **435**, 167714 (2023).

15. Tyler, W. J. & Murthy, V. N. Synaptic vesicles. *Current Biology* **14**, R294–R297 (2004).

16. Mori, A., Imai, Y. & Hattori, N. Lipids: Key Players That Modulate α-Synuclein Toxicity and Neurodegeneration in Parkinson's Disease. *Int J Mol Sci* **21**, 3301 (2020).

17. Mahul-Mellier, A.-L. *et al.* The process of Lewy body formation, rather than simply α-synuclein fibrillization, is one of the major drivers of neurodegeneration. *Proceedings of the National Academy of Sciences* **117**, 4971–4982 (2020).

18. Choi, Y. R., Park, S. J. & Park, S. M. Molecular events underlying the cell-to-cell transmission of α-synuclein. *FEBS J* **288**, 6593–6602 (2021).

19. McCormack, A. *et al.* Abundance of Synaptic Vesicle-Related Proteins in Alpha-Synuclein-Containing Protein Inclusions Suggests a Targeted Formation Mechanism. *Neurotox Res* **35**, 883–897 (2019).

20. Kiechle, M., Grozdanov, V. & Danzer, K. M. The Role of Lipids in the Initiation of α-Synuclein Misfolding. *Front Cell Dev Biol* **8**, (2020).

21. Meade, R. M., Fairlie, D. P. & Mason, J. M. Alpha-synuclein structure and Parkinson's disease – lessons and emerging principles. *Mol Neurodegener* **14**, 29 (2019).

22. Jao, C. C., Der-Sarkissian, A., Chen, J. & Langen, R. Structure of membrane-bound α-synuclein studied by site-directed spin labeling. *Proceedings of the National Academy of Sciences* **101**, 8331–8336 (2004).

23. Bartels, T. *et al.* The N-Terminus of the Intrinsically Disordered Protein α-Synuclein Triggers Membrane Binding and Helix Folding. *Biophys J* **99**, 2116–2124 (2010).

24. Kaur, U. & Lee, J. C. Unroofing site-specific α-synuclein–lipid interactions at the plasma membrane. *Proceedings of the National Academy of Sciences* **117**, 18977–18983 (2020).

25. Lee, H.-J., Choi, C. & Lee, S.-J. Membrane-bound α-Synuclein Has a High Aggregation Propensity and the Ability to Seed the Aggregation of the Cytosolic Form. *Journal of Biological Chemistry* **277**, 671–678 (2002).

26. Ohgita, T., Namba, N., Kono, H., Shimanouchi, T. & Saito, H. Mechanisms of enhanced aggregation and fibril formation of Parkinson's disease-related variants of α-synuclein. *Sci Rep* **12**, 6770 (2022).

27. Lima, V. de A., do Nascimento, L. A., Eliezer, D. & Follmer, C. Role of Parkinson's Disease-Linked Mutations and N-Terminal Acetylation on the Oligomerization of α-Synuclein Induced by 3,4-Dihydroxyphenylacetaldehyde. *ACS Chem Neurosci* **10**, 690–703 (2019).

28. Ysselstein, D. *et al.* Effects of impaired membrane interactions on α-synuclein aggregation and neurotoxicity. *Neurobiol Dis* **79**, 150–163 (2015).

29. Stöckl, M., Fischer, P., Wanker, E. & Herrmann, A. α-Synuclein Selectively Binds to Anionic Phospholipids Embedded in Liquid-Disordered Domains. *J Mol Biol* **375**, 1394–1404 (2008).

30. Sode, K., Ochiai, S., Kobayashi, N. & Usuzaka, E. Effect of Reparation of Repeat Sequences in the Human α-Synuclein on Fibrillation Ability. *Int J Biol Sci* 1–7 (2007) doi:10.7150/ijbs.3.1.

31. Chandra, S., Chen, X., Rizo, J., Jahn, R. & Südhof, T. C. A Broken α-Helix in Folded α-Synuclein. *Journal of Biological Chemistry* **278**, 15313–15318 (2003).

32. Rhoades, E., Ramlall, T. F., Webb, W. W. & Eliezer, D. Quantification of α-Synuclein Binding to Lipid Vesicles Using Fluorescence Correlation Spectroscopy. *Biophys J* **90**, 4692–4700 (2006).

33. Kim, S., Atwood, H. L. & Cooper, R. L. Assessing accurate sizes of synaptic vesicles in nerve terminals. *Brain Res* **877**, 209–217 (2000).

34. Binotti, B., Jahn, R. & Pérez-Lara, Á. An overview of the synaptic vesicle lipid composition. *Arch Biochem Biophys* **709**, 108966 (2021).

35. Takamori, S. *et al.* Molecular Anatomy of a Trafficking Organelle. *Cell* **127**, 831–846 (2006).

36. Middleton, E. R. & Rhoades, E. Effects of Curvature and Composition on α-Synuclein Binding to Lipid Vesicles. *Biophys J* **99**, 2279–2288 (2010).

37. Sarchione, A., Marchand, A., Taymans, J.-M. & Chartier-Harlin, M.-C. Alpha-Synuclein and Lipids: The Elephant in the Room? *Cells* **10**, 2452 (2021).

38. Alderson, T. R. & Markley, J. L. Biophysical characterization of α-synuclein and its controversial structure. *Intrinsically Disord Proteins* **1**, e26255 (2013).

39. Srinivasan, E. *et al.* Alpha-Synuclein Aggregation in Parkinson's Disease. *Front Med (Lausanne)* **8**, (2021).

40. Frieg, B. *et al.* The 3D structure of lipidic fibrils of α-synuclein. *Nat Commun* **13**, 6810 (2022).

41. Freeman, G. J., Casasnovas, J. M., Umetsu, D. T. & DeKruyff, R. H. *TIM* genes: a family of cell surface phosphatidylserine receptors that regulate innate and adaptive immunity. *Immunol Rev* **235**, 172–189 (2010).

42. Solinas, C., De Silva, P., Bron, D., Willard-Gallo, K. & Sangiolo, D. Significance of TIM3 expression in cancer: From biology to the clinic. *Semin Oncol* **46**, 372–379 (2019).

43. Liu, F., Liu, Y. & Chen, Z. Tim-3 expression and its role in hepatocellular carcinoma. *J Hematol Oncol* **11**, 126 (2018).

44. Kuchroo, V. K., Umetsu, D. T., DeKruyff, R. H. & Freeman, G. J. The TIM gene family: emerging roles in immunity and disease. *Nat Rev Immunol* **3**, 454–462 (2003).

45. DeKruyff, R. H. *et al.* T Cell/Transmembrane, Ig, and Mucin-3 Allelic Variants Differentially Recognize Phosphatidylserine and Mediate Phagocytosis of Apoptotic Cells. *The Journal of Immunology* **184**, 1918–1930 (2010).

46. Rivel, T., Ramseyer, C. & Yesylevskyy, S. The asymmetry of plasma membranes and their cholesterol content influence the uptake of cisplatin. *Sci Rep* **9**, 5627 (2019).

47. Suzuki, J. & Nagata, S. Phospholipid Scrambling on the Plasma Membrane. in 381–393 (2014). doi:10.1016/B978-0-12-417158-9.00015-7.

48. Wu, N. & Veillette, A. Lipid scrambling in immunology: why it is important. *Cell Mol Immunol* **20**, 1081–1083 (2023).

49. Tian, T. & Li, Z. Targeting Tim-3 in Cancer With Resistance to PD-1/PD-L1 Blockade. *Front Oncol* **11**, (2021).

50. Cong, Y., Liu, J., Chen, G. & Qiao, G. The Emerging Role of T-Cell Immunoglobulin Mucin-3 in Breast Cancer: A Promising Target For Immunotherapy. *Front Oncol* **11**, (2021).

51. Kim, H.-S. *et al.* Glial TIM-3 Modulates Immune Responses in the Brain Tumor Microenvironment. *Cancer Res* **80**, 1833–1845 (2020).

52. Smith, C. M., Li, A., Krishnamurthy, N. & Lemmon, M. A. Phosphatidylserine binding directly regulates TIM-3 function. *Biochemical Journal* **478**, 3331–3349 (2021).

53. Mishima, Y. *et al.* 235 Novel biparatopic TIM-3 antibody effectively blocks multiple inherent ligands and activates anti-tumor immunity. *J Immunother Cancer* **9**, A251–A251 (2021).

54. Pagliano, O. *et al.* Tim-3 mediates T cell trogocytosis to limit antitumor immunity. *Journal of Clinical Investigation* **132**, (2022).

55. Zhang, D. *et al.* Identification and characterization of M6903, an antagonistic anti–TIM-3 monoclonal antibody. *Oncoimmunology* **9**, (2020).

56. Kerr, D. *et al.* How Tim proteins differentially exploit membrane features to attain robust target sensitivity. *Biophys J* **120**, 4891–4902 (2021).

57. Tietjen, G. T. *et al.* Molecular mechanism for differential recognition of membrane phosphatidylserine by the immune regulatory receptor Tim4. *Proceedings of the National Academy of Sciences* **111**, (2014).

58. Weber, J. K. & Zhou, R. Phosphatidylserine-Induced Conformational Modulation of Immune Cell Exhaustion-Associated Receptor TIM3. *Sci Rep* **7**, 13579 (2017).

CHAPTER 3: Methodology

3.1. Introduction

While the overarching theme of this book is protein:membrane interactions, α-synuclein and the TIMs constitute two widely differing protein systems. To account for the diversity of the systems and differing goals in studying membrane binding, a wide set of protein expression systems, labeling methods, and lipid binding assay were used to quantify protein:membrane interactions. These assays include vesicle sedimentation, fluorescence polarization (FP), tryptophan fluorescence, ^{1}H-^{15}N Heteronuclear Single Quantum Coherence Nuclear Magnetic Resonance (^{1}H-^{15}N HSQC NMR), and circular dichroism (CD) spectroscopy. In addition to binding assays, the structural features of the hTIM3 system were further explored through the parallel use of X-ray Reflectivity (XR) and Molecular Dynamics (MD). The aim of this chapter is to discuss the methodology of the techniques to establish the fundamental principles that form the bases for analyses within **Chapters 4, 5,** and **6**.

3.2. Lipid-Binding Assays

A number of lipid-binding assays were used to quantify and characterize protein:lipid binding for α-synuclein and hTIM3. The variety of used techniques intends to cover the wide range of experimental parameters (e.g., protein concentration) and bound state identities (e.g., adsorbed vs. membrane-inserted) within the studied protein systems. The following section discusses the experimental design and methodology of these assays along with their experimental uses and limitations within the protein systems.

3.2.1. Vesicle Sedimentation Assay

3.2.1.1. Vesicle Sedimentation Assay: Methodology

A vesicle sedimentation assay can be used to identify buffer and lipid conditions under which a protein would bind to lipid membranes.[1] Briefly, sucrose-loaded large unilamellar vesicles (LUVs) are incubated with the protein sample, and then a sequence of centrifugation steps separates bound protein from unbound protein and the heavy liposomes, allowing for "bound" protein analysis through gel electrophoresis (full description of this method can be found in **Fig. 3.1**) While this method is not suitable for precise quantification, it does provide a binary or qualitative readout of protein:membrane binding. Noteworthy, this method cannot differentiate between bound and adsorbed states, and therefore it should be carefully used with protein samples that are susceptible to precipitation or under conditions with low ionic strength. Within this book, vesicle sedimentation was used to confirm the Ca^{2+} and PS-specificity of hTIM3's lipid binding properties (**Chapter 6**).

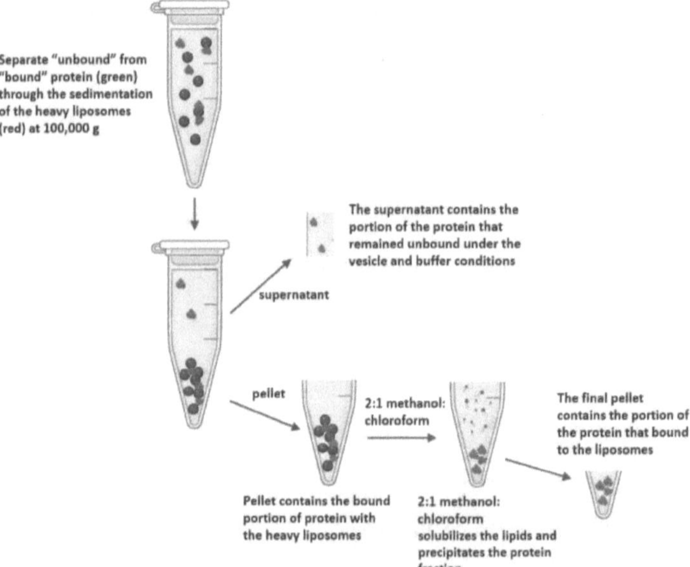

Fig. 3.1. Schematic of vesicle sedimentation assay.

Fig. 3.1, continued. Schematic of the methodology of the vesicle sedimentation assay, where sucrose-loaded "heavy" liposomes (red) are used to separate proteins (green) that are lipid membrane-bound from the unbound ones. Prior to analysis of the bound and unbound fractions through gel electrophoresis, the bound fraction of the protein is separated from the liposomes by solubilizing the lipids and precipitating the protein in a 2:1 methanol: chloroform solution.

3.2.1.2. Vesicle Sedimentation Assay: Protocol

Sucrose-loaded LUVs at a concentration of 2 mM were prepared as described in **Chapter 3.4.5.2.**

except lipid sample were resuspended in 40 mM HEPES, 100 mM NaCl, and 0.5 M sucrose, pH

7.2 in place of HBS. Vesicles were diluted 5x into Buffer A (40 mM HEPES, 100 mM NaCl, pH

7.2), and then heavy liposomes were isolated by centrifugation at 100,000g for 30 min at 4 °C. The

pellet was transferred to a microfuge tube by 3x washes with 50 μL of HBS (final volume 150 μL),

and then centrifuged at 13,000 rpm for 10 minutes at 4 °C. The final pellet was then resuspended

in 250 μL HBS (final concentration ~10 mM), and the heavy liposomes were incubated with

protein and rotated at 4 °C for 30 minutes. Samples were then centrifuged at 13,000 rpm for 30

minutes at 4 °C to separate bound and unbound fractions. The pellet (containing "bound" protein

and the heavy liposomes) was resuspended in 25 μL of 2:1, v/v methanol:chloroform to solvate

the liposomes and precipitate the protein. The sample was then centrifuged at 13,000 rpm for 15

minutes at 4 °C, and the pellet, containing the bound fraction of protein, was resuspended in HBS

buffer. All final bound and unbound fractions were analyzed through gel electrophoresis.

3.2.2. Fluorescence Polarization

3.2.2.1. Fluorescence Polarization: Methodology

Fluorescence polarization (FP) can monitor protein:lipid binding by leveraging the relationship

between the rotational speed of a fluorophore and depolarization of its emission (**Fig. 3.2**)[2]. Briefly,

a fluorophore is excited with polarized light, and then the emission intensity is measured in the parallel ($I_{\parallel}$) and perpendicular ($I_{\perp}$) directions, which can then be used to calculate an anisotropy (A) value (**Eq. 3.1**).

$$A = \frac{I_{\parallel} - I_{\perp}}{I_{\parallel} + 2\,I_{\perp}}$$

Eq. 3.1

A fully immobile fluorophore would emit light corresponding to the angle (θ_d) between the absorbing dipole and emitting dipole (**Eq. 3.2**), resulting in a maximum anisotropy value of 0.4.

$$A = \frac{3\cos^2(\theta_d) - 1}{5}$$

Eq. 3.2

However, the emission of a tumbling fluorophore would be depolarized, resulting in a lower measured $I_{\parallel}$ and a higher $I_{\perp}$ (**Fig. 3.2b,c**). As the tumbling speed of an object is correlated with its molecular weight, the smaller "unbound" state would have a smaller anisotropy (A_{min}) than the anisotropy (A_{max}) of the larger complex that is formed in the "bound state" (**Fig. 3.2a**). Generally, FP assays tag the ligand with a fluorophore, as this leads to a larger change in the polarization of the emission spectrum upon binding to the larger protein.[2] However, within our assay, we tag the protein, since the protein is much smaller than its binding partner, the lipid vesicles.

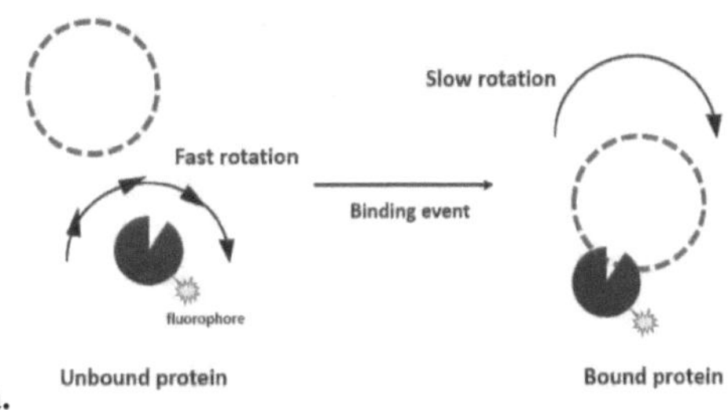

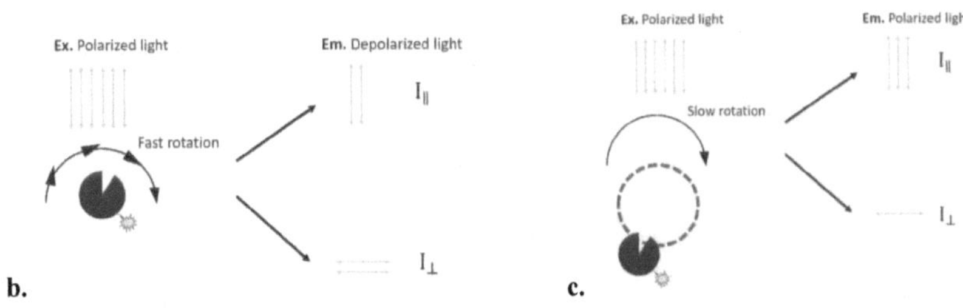

Fig. 3.2. Schematic of fluorescence polarization in monitoring protein binding to lipid vesicles. **a.** Tumbling speed of protein decreases after binding to the larger lipid vesicles. **b and c.** Polarization states of excited (Ex.) and emitted light (Em.) of fluorophore-tagged protein unbound **b.** and bound **c.** to lipid vesicles.

Using a linear combination of the anisotropy values of the bound and unbound states, the fraction of protein bound (f_{bound}) at intermediate lipid concentrations can be calculated (**Eq. 3.3**), which in turn allows for the construction of a binding curve.

$$A_{measured} = A_{min} * (1 - f_{bound}) + A_{max} * f_{bound} \qquad \text{Eq. 3.3}$$

As this method relies on differences in the tumbling speed between the fluorophore's bound and unbound states, FP largely probes only at association between the protein and the lipid vesicle and is therefore limited in its ability to distinguish between different bound states. Within the work shown here, FP was used to quantify lipid-binding of α-synuclein (further discussed in **Chapter 5**), where the primary focus of our work was to validate the technique for future high-throughput lipid-binding experiments with α-synuclein. We additionally attempted to use FP to quantify hTIM3:membrane binding, but we were unable to successfully circumvent the limitations of disulfide bonds in the TIM proteins. As cysteine-labeling is not possible (as it would disrupt protein fold), we attempted to exploit the lower pKa value of the N-terminal amine group of the protein for fluorophore labeling.[3] While we were able to monitor protein:membrane binding for mTIM3, binding appeared to be significantly reduced in comparison to values found from tryptophan fluorescence binding assays, likely due to the partial labeling of critical lysine residues that would impede binding.

3.2.2.2. *Fluorescence Polarization: Protocol*

Wells for FP measurements were assembled on a Greiner CELLSTAR® 96 well plate with black polystyrene wells and a micro-clear bottom using a total volume of 150 μL across all samples. FP measurements were taken on the BioTek Synergy Neo2, using filter sets "Dual FP 4" and "FP 485/530 LUM" for excitation and emission measurements of Alexa488-tagged α-synuclein. We found that a protein concentration of 50 nM yielded the best signal:noise ratio, and lipid concentrations were adjusted to match the protein:lipid ratio of tryptophan fluorescence experiments (**Chapter 5.3**).

3.2.3. Tryptophan Fluorescence Spectroscopy

3.2.3.1. *Tryptophan Fluorescence Spectroscopy: Methodology*

Tryptophan is a naturally occurring fluorescent amino acid with an emission spectrum that reports on its environment.[4] While the residue often occupies an internal hydrophobic core within protein structure, external tryptophan residues can dictate protein binding by forming hydrophobic contacts with its ligands. Several proteins, such as members of the TIM family[5] and milk fat globule-EGF factor 8 protein (MFGE8), another PS-binding protein,[6] insert a tryptophan residue into the hydrophobic core of lipid membranes as a part of the protein's membrane-bound state (**Fig. 3.3a**). The transition from the aqueous external environment to the hydrophobic environment of the membrane results in a change in the tryptophan's emission spectrum, and it can therefore be used to monitor protein binding.

Excitation of tryptophan at a wavelength of 280 nm yields an emission spectrum in the 300-420 nm range that is comprised of emission from two states: ^1L_a and ^1L_b, where the ^1L_a state dominates in polar environments.[7] With increased hydrophobicity in the environment, tryptophan emission blue shifts towards the ^1L_b state and the overall emission spectrum experiences less quenching (higher fluorescence quantum yield) from surrounding water molecules.[8] Altogether, the insertion of the tryptophan into the hydrophobic membrane results in a blue shift and an increase in intensity for the emission spectrum (**Fig. 3.3b**). Using a linear combination of the spectra corresponding to the "bound" (S_{bound}) and "unbound" states ($S_{unbound}$) from 300 nm to 420 nm, we can determine the fraction of tryptophan bound (f_{bound}) at intermediate lipid concentrations to construct a lipid-binding curve[5] (**Eq. 3.4**).

$$S_{measured} = f_{bound} * S_{bound} + (1 - f_{bound}) * S_{unbound} \qquad \textbf{Eq. 3.4}$$

It is worth noting that the observed fully bound protein (b_{max}) is close but not identical to the total protein concentration (b_{total}), and therefore the true S_{bound} is estimated from the saturated, maximally bound spectrum (S_{max}) (**Eq. 3.5**).

$$S_{max} = \frac{b_{max}}{b_{total}} * S_{bound} + \left(1 - \frac{b_{max}}{b_{total}}\right) * S_{unbound} \qquad \textbf{Eq. 3.5}$$

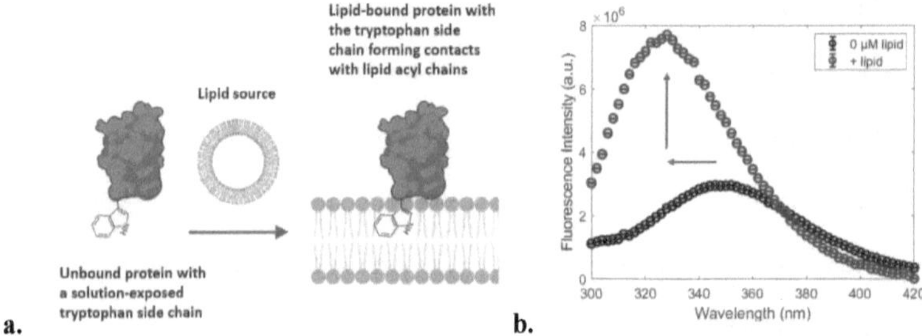

Fig. 3.3. Schematic of tryptophan fluorescence in lipid-binding. **a.** Solution-exposed side chain of a tryptophan residue of an unbound state (left) and one burying into the lipid membrane and forming contacts with lipid acyl chains when bound (right). **b.** Representative tryptophan emission spectra of unbound (black) and bound (blue) states of a protein that inserts a tryptophan residue into the lipid membrane upon binding.

As changes to tryptophan fluorescence occur when the tryptophan residue interacts with the acyl changes of the lipid bilayer, this method is site-specific to tryptophan-inserted states. In comparison with methods like FP, tryptophan fluorescence can differentiate between bound states, given the sufficient difference in the tryptophan emission spectrum between the states. However, this method has several limitations. First and foremost, the protein of interest must contain a tryptophan that undergoes a change in its environment upon binding, and WT α-synuclein does not. To overcome this limitation, we expressed protein with the F4W and F94W mutations, as phenylalanine is a similarly hydrophobic amino acid and has been previously validated to not affect the affinity of α-synuclein for lipid membranes.[9] These residues are located on the hydrophobic side of the amphipathic helical wheel for α-synuclein and the site specificity of this method allows us to independently probe at the bound state of the two helices in the protein. **Chapter 4** contains a more thorough discussion of the utility of the site-specificity of the method in characterizing the bound states of α-synuclein.

3.2.3.2. Tryptophan Fluorescence Spectroscopy: Protocol

Tryptophan fluorescence experiments were taken on a Horiba Fluorolog-2 spectrophotometer (Horiba, Kyoto, Japan) using a USHIO Xenon short arc lamp with the sample in a 10-mm path-length quartz cuvette (Starna Cells, Inc., Atascadero, CA). All samples were excited at a wavelength of 280 nm, and the emission spectrum was measured from 300–420 nm (slid width of 2 nm). Samples were made for protein concentrations of 125 nM in HBS at 37 °C at varying concentrations of lipid (0-2500 μM). Prior to measurement, samples were allowed to equilibrate for 1.5 minutes with stirring. For each sample, a lipid-only scan was taken as the background spectrum to account for lipid scattering. For each lipid concentration, a minimum of three replicates were collected.

3.2.4. Circular Dichroism Spectroscopy

3.2.4.1. Circular Dichroism Spectroscopy: Methodology

Circular Dichroism (CD) is a spectroscopy-based method that can be used to monitor changes to the secondary structure of protein samples.[10,11] In short, the method relies on a chiral molecule's differences in absorption of left and right circularly polarized light (LCP and RCP), where the ellipticity of the polarization (θ_e) can be written as:

$$\tan \theta_e = \frac{E_R - E_L}{E_R + E_L} \qquad \text{Eq. 3.6}$$

with E_R and E_L being the magnitudes of the electric field vectors of RCP and LCP. From this, we can define the molar ellipticity, $[\theta_e]$ in units of deg·cm^2·dmol^{-1} as follows:

$$\theta_e = 3298.2 \, \Delta\varepsilon \qquad \text{Eq. 3.7}$$

where $\Delta\varepsilon$ is the molar circular dichroism value. Differences in absorption of LCP and RCP are dependent on both the wavelength of the polarized light and the structure of the measured sample

(**Fig. 3.4**). Therefore, CD measurements can be used to 1) identify structural features of a protein, and 2) relate differences in sample conditions (e.g., buffer, ligand, etc.) to structural changes.

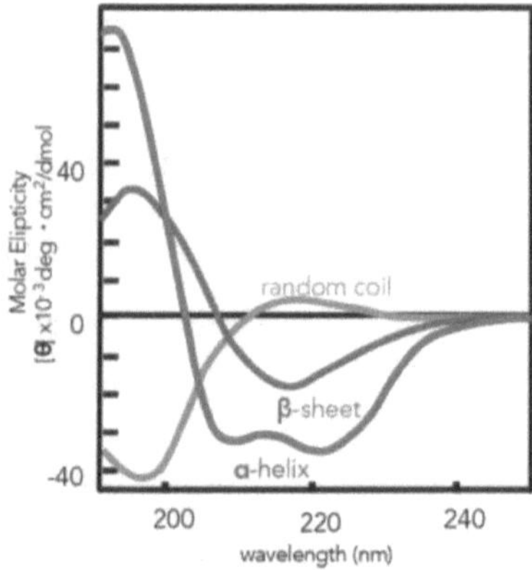

Fig. 3.4. Theoretical CD spectra of various protein structures. Figure adapted from Greenfield *et al.*[10]

Within our work, we use CD spectra to examine changes to the structure of α-synuclein with the addition of lipid membranes, as α-synuclein is known to convert from a disordered protein to one with more α-helical character upon binding (further detailed in **Chapter 2.2.1**). Using these changes, we can quantify protein:lipid binding and observe changes to bound states of the protein. Of note, this method requires μM protein concentrations, and therefore cannot be used as a direct comparison to the nM protein concentrations of tryptophan fluorescence measurements for α-synuclein.

3.2.4.2. Circular Dichroism Spectroscopy: Protocol

All CD measurements were taken on the CD Spectrophotometer Jasco J-1500 using the default settings of the instrument in a 1-mm path length quartz cuvette in HBS buffer at 25 °C.

3.2.5. ¹H-¹⁵N HSQC NMR

3.2.5.1. ¹H-¹⁵N HSQC NMR: *Methodology*

^{1}H-^{15}N Heteronuclear Single Quantum Coherence Nuclear Magnetic Resonance (HSQC NMR) is a method that can monitor protein environment[12] and therefore be used to quantify protein:lipid binding.[13] In brief, the method measures the transfer of magnetization between a proton to the nucleus of its bonded ^{15}N, yielding a peak on a two-dimensional spectrum of the chemical shifts of the ^{1}H and the ^{15}N (**Fig. 3.5**). However, upon binding to larger lipid vesicles this signal broadens and becomes undetectable due to the slower tumbling speed of the complex in comparison to free protein.[14] This property of the measurements can be used to quantify the amount of free vs. bound protein through the integration of residues' peak volumes.

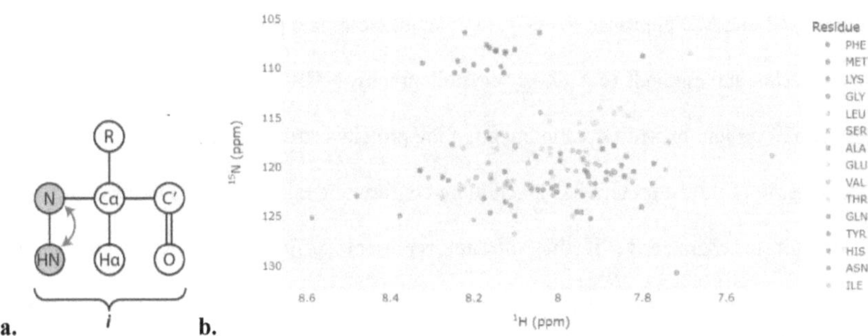

Fig. 3.5. a. Magnetization transfer (red arrow) between the proton and nitrogen in the backbone of a protein. **b.** Sample ^{1}H-^{15}N HSQC NMR spectrum of α-synuclein, BMRB entry: 51148[15]

3.2.5.2. ^{1}H-^{15}N HSQC NMR: *Protocol*

Within our work here, we use ^{1}H-^{15}N HSQC NMR to quantify the binding of α-synuclein to lipid membranes[16] (**Chapter 5.5**). Measurements were taken in the Biomolecular NMR facility with the help of Joseph Sachleben on a Bruker AVANCE IIIHD 600 NMR spectrometer. Samples with a total volume of 500 μL were made in HBS buffer with an added 50 μL of D$_2$O.

3.3. Structural Interrogation through XR and MD

3.3.1. Paralleled Use of XR and MD: Introduction

The structure of peripheral membrane-bound protein states cannot be determined through traditional structural methods like X-ray crystallography and cryo-electron microscopy (cryo-EM) as these methods are best suited for solution-based protein samples.[17] X-ray reflectivity (XR) is a method that retains a protein in its native, membrane-bound environment and can overcome some of these limitations to provide electron density information perpendicular to the membrane on the lipid membrane-bound states.[18] However, it is difficult to isolate a structure with high resolution solely from XR and the use of other methods in parallel, such as molecular dynamics (MD), can aid in the characterization of a protein's conformational state at the membrane and identify protein:membrane contacts.[19] Here we discuss the methodology, uses, and limitations of a combined XR and MD approach to analyzing membrane-bound protein systems.

Many experimental methods (e.g., X-ray crystallography, NMR, cryo-EM) used to determine the structure of a protein provide a static image of the protein's structure in solution or in its ligand-bound state. In reality, a protein has much more conformational flexibility and these fluctuations are difficult to characterize if they do not represent sufficiently differing protein states. Furthermore, these methods are best for characterizing the more rigid portions of protein and often fail to provide resolution for the more flexible portions of proteins such as loop or disordered regions.

Some of the experimental limitations of these methods are linked to the conditions under which data can be collected. For example, X-ray crystallography can only characterize systems that are amenable to forming rigid, repeating arrays, while cryo-EM requires the freezing of a protein sample,[20] destructive to certain protein states. Furthermore, X-ray crystallography and cryo-EM provide only a static image of a protein's structure, and extending a method like cryo-EM to capturing a protein transition would require the collection of many sample images. While NMR-based methods can retain the protein sample in buffer (and thus, a more native-like state) and provide dynamic, structural information for conformational changes, these methods can be prohibitively expensive for proteins that can only be made by insect- or mammalian- expression systems due to the necessary ^{13}C- and ^{15}N- labeling.

Structural-determination of membrane-bound states is even more challenging due to the added complexity and system flexibility of a lipid membrane. Along with global protein structure, characterizing a protein:membrane systems requires information on the protein's position with respect to the lipid membrane, where lipid:protein contacts are not static and can often be interchangeable and at the same time still influencing protein orientation and insertion depth. While X-ray crystallography can capture protein states in complex with single lipid molecules, crystallizing a state that retains the native membrane environment is nearly impossible. Cryo-EM is better suited for analyzing membrane-bound states but is limited to static images of the protein. A combined approach through XR and MD[21] attempts to address the above limitations through conditions that reflect the full conformational space of *in vivo* membrane-bound protein states. XR measurements of the protein:membrane system are taken in solution with relevant protein:lipid contacts intact. The electron density information from XR measurements can then be fit with potential bound orientation and conformations identified by MD simulations, which can further inform on the stability and flexibility of the identified states. Altogether, the combined approach aims to provide a broader, comprehensive view of proteins at lipid membranes.

3.3.2. Experimental Setup for XR Measurements

The primary benefit to the experimental set-up for XR measurements is the ability to retain the protein:lipid system in a more native-like environment. Within this system, the protein binds to the lipid headgroups region of the monolayer, which mimics the features of a bilayer leaflet and is therefore a suitable model for many peripheral membrane proteins.[22] However, the flat monolayer model is likely not suitable for lipid:protein systems that span both leaflets, such as transmembrane proteins, or ones that are sensitive to other membrane features, such as high curvature. Briefly, a monolayer of lipids is deposited onto a buffer subphase in a Langmuir trough, where it is then compressed to the desired surface pressure by a movable barrier to mimic lipid packing density in membranes. The surface pressure is determined from the Wilhelmy plate, a piece of filter paper that is immersed into the buffer and directly measures the downward force exerted by the buffer meniscus. Thus, the surface pressure control of the Langmuir trough can be used to better mimic the spacing and compression of the lipid molecules in a physiologically-relevant bilayer leaflet.[22] Protein is then injected under the surface of the monolayer and allowed to equilibrate to its lipid-bound state (**Fig. 3.6a**). Prior to XR data collection, the sample chamber is flushed with helium to reduce oxidative damage. XR measurements are then taken at specular reflection with varying incident angles (θ_i) of the incoming beam to yield a reflectivity curve that plots the normalized reflectivity (**R/R_F**) against the momentum transfer along the z direction (**Q_z**) (**Fig. 3.6b**, the theoretical framework for these terms are further covered in **Section 3.3.3.1**).This set-up allows for experimental control over lipid and buffer (e.g., Ca^{2+} concentrations) conditions, and protein measurements are taken in the presence of non-static lipid:protein contacts.

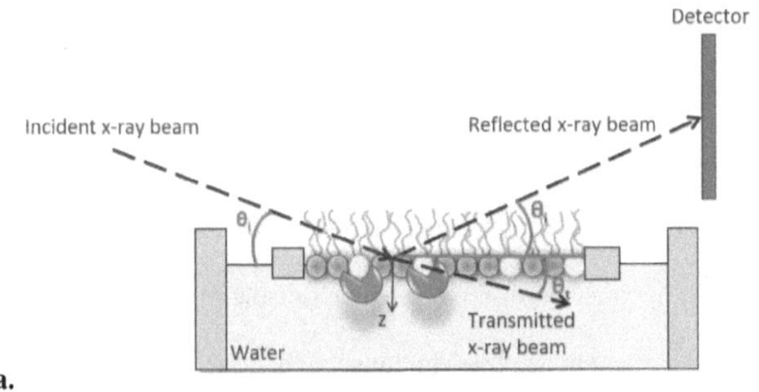

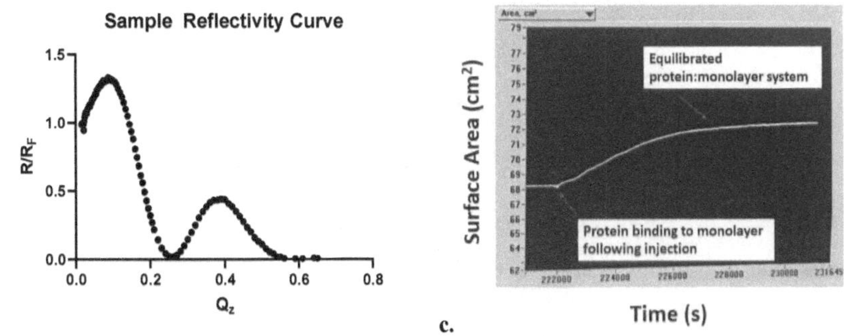

Fig. 3.6. a. Langmuir trough set-up for XR measurements of protein (orange) bound to a mixed lipid monolayer (with blue and yellow headgroups). **b.** Representative reflectivity curve of an XR scan, plotting the normalized reflectivity (***R/R_F***) against the momentum transfer along the *z* direction (***Q_z***). **c.** Representative surface area measurements following a binding event under constant pressure-control.

Protein binding and equilibration can be monitored through the surface pressure of the monolayer, as the insertion of the protein into the membrane often increases the effective pressure, and therefore to maintain constant pressure (Π), the Langmuir trough must expand its surface area (A) (**Fig. 3.6c**). Generally, the monolayer is compressed and maintained at surface pressures within the range of 20-30 mN/m to mimic the equivalent surface pressures of cell membranes (25-35 mN/m[23]). Through lower surface pressure, we can accelerate equilibration between the protein and lipids by increasing lipid spacing and decreasing the required energy to displace lipids to accommodate protein residues within the membrane.

Of note, a monolayer does not fully reflect the internal characteristics of a bilayer structure of lipid membranes and this experimental set-up may not be suitable for protein systems that insert deep into a bilayer (e.g., transmembrane proteins). In the monolayer, the hydrophilic lipid headgroups face the aqueous subphase and the hydrophobic lipid acyl chains extend above the air-water interface of the buffer. Ongoing efforts aim to extend XR-based protein measurements to bilayer systems. However, for proteins such as hTIM3 that minimally insert past the lipid headgroup region, a monolayer-bound state can approximate the bilayer-bound state and is suitable for our system.[18]

3.3.3. Analysis of XR Data

To determine a protein's structure and orientation from an XR curve, we must simulate the expected reflectivity from models of the electron density of the monolayer and the protein. The following section discusses the fundamental principles underlying XR and their integration with our system using the Parratt Method.[24]

3.3.3.1. XR Analysis: Fundamentals of XR

At a single interface, an incoming x-ray beam undergoes reflection and transmission events that are dependent on the electron density of the interface. An incoming wavevector $\vec{k_1}$ at an incident angle of θ_I would be paired with a reflected wavevector of $\vec{k_R}$ at an angle of $\theta_R = \theta_I$ and a transmitted wavevector of $\vec{k_T}$ at an angle of θ_T (**Fig. 3.7a,b**). The transmission event is governed by Snell's Law, where the index of refraction n_j is dependent on the dispersion δ_j and absorption β_j of the medium (**Eq. 3.8**).

$$n_j = 1 - \delta_j + i\beta_j \qquad\qquad \textbf{Eq. 3.8}$$

While β_j is negligible within our system, the dispersion δ_j can be approximated by the average electron density $< \rho(z) >$ (**Eq. 3.9**). Therefore, the measured reflected beam will be directly related to the electron density of the interface.

$$\delta(z) = \frac{2\pi r_e}{|\vec{k_i}|^2} < \rho(z) > \qquad\qquad \textbf{Eq. 3.9}$$

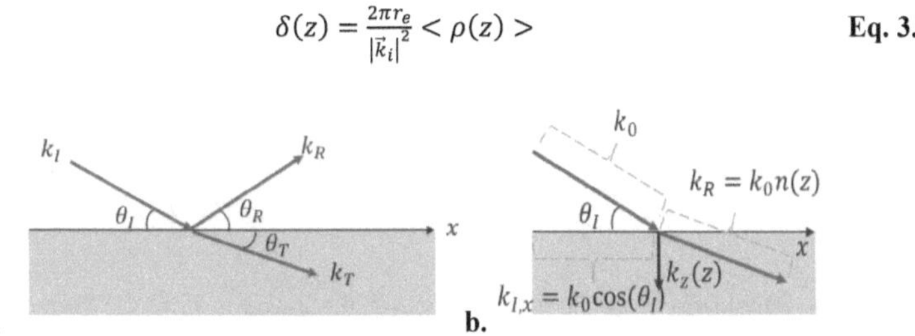

Fig. 3.7. a. Angles corresponding to transmission and reflection events at an interface for an incoming incident x-ray beam. **b.** Geometry of incident and transmitted wavevectors.

The reflected and transmitted light at a single interface can be described by the Fresnel coefficients $r(Q)$ and $t(Q)$ as a function of the momentum transfer vector Q and Q_T, given by the z component of the incident and transmitted wavevectors, respectively (**Eq. 3.10a,b**). These vectors are dependent on the incident angle θ_I (**Eq. 3.11**).

$$r(Q) = \frac{Q - Q_T}{Q + Q_T} \qquad\qquad \textbf{Eq. 3.10a}$$

$$t(Q) = \frac{2Q}{Q + Q_T} \qquad\qquad \textbf{Eq. 3.10b}$$

$$Q = \frac{4\pi \sin(\theta_I)}{\lambda} \qquad\qquad \textbf{Eq. 3.11}$$

While these equations describe the reflectivity and transmission events at a single interface, the experimentally measured curve is the result of many of these events down the z direction of the sample with respect to the incoming beam. We can extend **Eqs. 3.10, 3.11** to describe these events using the Parratt Method.[24] In short, the Parratt Method stratifies a sample into J distinct layers, where each layer has a thickness of d_j and an index of refraction of n_j (**Fig. 3.8**).

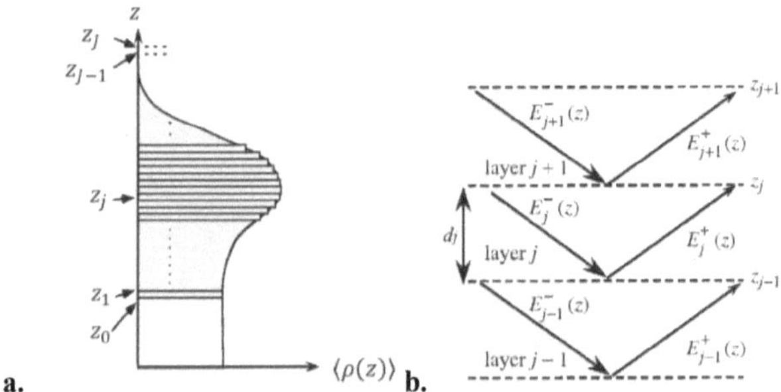

Fig. 3.8. a. Schematic of the stratification down the z axis used in the Parratt Method. Sample layers z_0 through z_j each have their own index of refraction n_j. **b.** Schematic of the propagation of the transmission and reflection events for the electric fields in each layer.

Dependent on the electron densities of layers j and j-1, the reflectivity and transmission events can be described through their coefficients (**Eq. 3.12a,b**):

$$r_{j,j-1} = \frac{k_{j,z} - k_{j-1,z}}{k_{j,z} + k_{j-1,z}} = -r_{j-1,j} \qquad \text{Eq. 3.12a}$$

$$t_{j,j-1} = \frac{2k_{j,z}}{k_{j,z} + k_{j-1,z}} = 1 + r_{j,j-1} \qquad \text{Eq. 3.12b}$$

For each layer, we can then denote the electric field, E_j, propagating down (-) or up (+) with respect to the z direction of the sample as follows, where A_j and B_j are the incident and reflected amplitudes, respectively:

$$E_j(z) = \begin{bmatrix} E^-_j(z) \\ E^+_j(z) \end{bmatrix} = \begin{bmatrix} A_j e^{-ik_{j,z} z} \\ B_j e^{ik_{j,z} z} \end{bmatrix} \qquad \text{Eq. 3.13}$$

We can describe the propagation of the incident and reflected waves through the layers as $P_j E_j(z_j)$, where P_j is used to inform subsequent layer's differences in amplitude and phase. With the interface matrix $I_{j-1,j}$, which satisfies the boundary conditions of layers j and j-1, we can write

the propagation of the electric field from layer j to j-1:

$$E_{j-1}(z_{j-1}) = I_{j-1,j}\, P_j E_j(z_j) \qquad\qquad \text{Eq. 3.14}$$

The contribution to the reflectivity at layer j can then be expressed as the total product of

propagation events from the bottommost layer 0 (**Eq. 3.15**):

$$\begin{bmatrix} E_0^-(z_0) \\ E_0^+(z_0) \end{bmatrix} = \mathbf{M} \begin{bmatrix} E_{j+1}^-(z_{j+1}) \\ E_{j+1}^+(z_{j+1}) \end{bmatrix} \quad \text{where} \quad \mathbf{M} = I_{0,1} P_1 I_{1,2} P_2 \dots P_j I_{j,j+1} = \begin{bmatrix} M_{11} & M_{12} \\ M_{21} & M_{22} \end{bmatrix} \text{Eq. 3.15}$$

The final reflectivity, R, can be given by:

$$R = \left| \frac{M_{21}}{M_{22}} \right|^2 \qquad\qquad \text{Eq. 3.16}$$

The principles and the Parratt Method described above provide a way to approximate the

reflectivity given the electron density of the stratified layers. To apply these models to our system,

we use electron density models of the lipid monolayer and the protein.

3.3.3.2. XR Analysis: Electron Density Models of the Monolayer and the Protein

The subphase, lipid monolayer, and the protein all contribute to the effective electron density

profile at the interface that affects the measured reflectivity curve. To account for the electron

densities of the lipid monolayer, we use the Slab Model[25] to discretize the lipid into the lipid head

and lipid tail, each given a constant electron density ρ_i and a thickness of L (**Fig. 3.9a**). At the

borders of the layers, thermal fluctuations smear electron density with an approximate roughness

of $\sigma = 3.4$ Å, a value that has been previously determined to minimize error in electron density

fitting for this system.[26] Of note, the electron density contribution of the helium in the superphrase

is negligible, $\rho_{air} = 0$ Å^{-3}, Therefore, the average electron density of the monolayer in our system can be given by **Eq. 3.17**, where the electron density of the subphase is approximately $\rho_{buffer} = 0.334\ \text{Å}^{-3}$.

$$\langle \rho(z) \rangle = \frac{\rho_{air} + \rho_{buffer}}{2} + \frac{\rho_{air} - \rho_{tail}}{2}\ \text{erf}\left(\frac{z}{\sqrt{2\sigma^2}}\right) + \frac{\rho_{tail} - \rho_{head}}{2}\ \text{erf}\left(\frac{z + L_{tail}}{\sqrt{2\sigma^2}}\right) +$$

$$\frac{\rho_{head} - \rho_{buffer}}{2}\ \text{erf}\left(\frac{z + L_{tail} + L_{head}}{\sqrt{2\sigma^2}}\right) \qquad\qquad \textbf{Eq. 3.17}$$

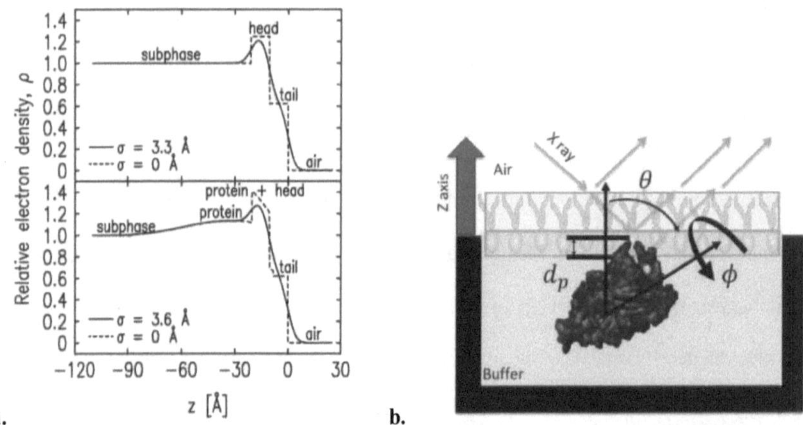

Fig. 3.9. a. Slab Model representation of the interfacial profile of a lipid-only system (top) and with added protein (bottom), where dashed lines demonstrate the delineation of electron density between the head and tail of the lipid when the roughness σ = 0. Figure adapted from *Málková et al.*[25] **b.** Representation of the Euler angles, θ and φ, and protein depth d_p of a protein bound to a lipid membrane. Figure adapted from Daniel Kerr.[21]

The electron density profile down the z axis of a protein placed at the monolayer will depend on its orientation, given by its Euler angles θ and φ, and depth of insertion d_p (**Fig. 3.9b**)[21]. For any given orientation, we can use a rotational matrix $R_{\theta,\phi}$ to transform the initial coordinates of each atom a_i, which are assigned from an existing model (e.g., crystal structure), to yield the new coordinates a_f. (**Eq. 3.18**). Of note, a protein structure that can be used to assign atom coordinates is a prerequisite for the use of this method, as coordinates cannot be assigned from protein sequence alone.

$$\vec{a_f} = \vec{a_i} * R_{\theta,\phi} = \begin{bmatrix} 1 & 0 & 0 \\ 0 & \cos\theta & -\sin\theta \\ 0 & \sin\theta & \cos\theta \end{bmatrix} \begin{bmatrix} \cos\phi & -\sin\phi & 0 \\ \sin\phi & \cos\phi & 0 \\ 0 & 0 & 1 \end{bmatrix} \vec{a_i} \qquad \textbf{Eq. 3.18}$$

To model each atom's contribution to the electron density profile, we use a Heaviside step function, H, where the electron density is modeled as n_i electrons within a van der Waal's sphere of radius r_i and an electron density of zero outside the sphere (**Eq. 3.19**).

$$\rho_i(\vec{r}) = \frac{n_i}{\frac{4}{3}*\pi r_i^3}\left[1 - H\left(\left|\vec{r} - \vec{a_f}\right|^2 - r_i^2\right)\right] \qquad \textbf{Eq. 3.19}$$

To approximate the electron density along the z axis and for its use within the subsequent analyses, we can "slice" the protein along the z direction to yield (**Eq. 3.20**):

$$\langle\rho_i(z)\rangle = \frac{n_i}{\frac{4}{3}*\pi r_i^3}\left[1 - H\left(\left|z - a_{f,z}\right|^2 - r_i^2\right)\right] * \left[\frac{\pi\left(r_i^2-(z-a_{f,z})^2\right)}{\pi r_i^2}\right] \qquad \textbf{Eq. 3.20}$$

Finally, we can account for electron density contributions from the buffer (due to space that is unoccupied by protein) by integrating over an area A_{rect} using predefined x and y lengths (**Eqs. 3.21, 3.22**). Thus, we can approximate the average electron density of the protein using grid spacings of Δx, Δy, and Δz, where spacings of 0.5 Å are sufficient for convergence of fit results[21].

$$\langle\rho(z)\rangle_{total} = \frac{\iint_{A_{rect}} \sum_i^N \rho_i(\vec{r})\, dx\, dy}{\iint_{A_{rect}} dx\, dy} \qquad \textbf{Eq. 3.21}$$

$$S_{rect} = \left(\max_i(a_{f,x} + r_i) - \min_i(a_{f,x} - r_i)\right) \times \left(\max_i(a_{f,y} + r_i) - \min_i(a_{f,y} - r_i)\right) \qquad \textbf{Eq. 3.22}$$

The principles described within **Sections 3.3.3.1.** and **3.3.3.2.** provide a method to approximate the electron density of a lipid monolayer:protein system, which can then be used to simulate a reflectivity curve for a particular protein orientation and depth of insertion. By comparing predicted reflectivity curves of various protein models at various orientations and insertion depths to experimentally-obtained reflectivity curves, we can identify probable bound states of the protein at the membrane.[27] While not necessary for all protein systems, the work within this book uses MD to expand the search space of possible bound states and orientations from published crystal structures of hTIM3. The following section briefly discusses the methodology behind MD simulations and its use within the fitting of reflectivity data, but a more thorough description of the integration of these two methods can be found in **Chapter 6** within the analysis of hTIM3 reflectivity data.

3.3.3.3. XR: Protocol

All x-ray reflectivity experiments were performed at ChemMatCARS sector 15ID the Advanced Photon Source at Argonne National Laboratory using a custom-built Langmuir trough.[26] All materials were acid-cleaned prior to use, and care was taken to ensure minimal contamination from dust or other surface-active agents. The trough was first thoroughly cleaned with methanol, acetone, and chloroform followed by a water rise. Following cleaning, a stir bar and ~55 mL of buffer were added to the trough and compression test to eliminate surface contaminants was performed to ensure that surface pressure remained at $\Delta\Pi = 0$ mN/m from a fully expanded to a fully compressed trough. Lipids stock in chloroform was then deposited onto the buffer surface, the chloroform was allowed to evaporate for ~2 minutes, and the chamber was flushed with helium until the oxygen level was less than 2%. The monolayer was then compressed to a surface pressure of 2-3 mN/m, and a reflectivity scan was taken of the lipid-only system θ.

Following the lipid-only scan, protein, preincubated with 2-4 mM Ca^{2+}, was injected under the surface of the lipid film and equilibration with the monolayer was monitored through the expansion of the trough's area under constant pressure control. Of note, preincubation with Ca^{2+} must be done immediately prior to injection, as long-term preincubation could lead to protein aggregation (presumably due to Ca^{2+}-binding pocket opening and exposure of hydrophobic residues). Additionally, injection of hTIM3 must be carefully done to avoid bubbles and disruption to the monolayer, thus eliminating seeding protein aggregation (see **Chapter 6** for conditions that optimized hTIM3 binding). Following equilibration, 3-4 repeat scans were taken of the lipid monolayer:protein system at varying x positions along the trough.

Reflectivity scans were all taken for Q_z range of 0.016-0.65 $Å^{-1}$ with an x-ray wavelength of ~1.24 Å. Detector optimization and raw data processing for our experiments were done with the help of and code developed by Wei Bu and Mati Meron at NSF's ChemMatCARS beamline at Argonne. Protein fits, based on the principles in the section above, for X-ray reflectivity curves were analyzed using custom MATLAB code built by Daniel Kerr and Gregory Tietjen[5].

3.3.4. Molecular Dynamics Simulations

3.3.4.1. MD Simulations: Introduction

Molecular dynamics (MD) simulations can supplement experimental methods like XR by providing a theoretically based method to compare potential protein conformations and bound states. While protein fits of reflectivity data can be accomplished with published crystal structures, MD-generated states can further refine the fit and identify changes that occur upon protein binding. Additionally, MD can provide insight into the mechanisms behind protein conformational changes upon membrane-engagement and corresponding protein:lipid contacts. The following section discusses the methodology behind MD in exploring membrane-bound states and their use within fitting of reflectivity data.

3.3.4.2. MD Simulations: HMMM Model

MD can be used to capture and characterize peripheral membrane binding events for smaller protein systems. However, all-atom simulations of lipid membranes can be resource-expensive, as the slow diffusion of lipid molecules increases simulation time required for the protein to explore its conformational space and approach an equilibrated state. The slow diffusion of the lipids renders it impossible to use all-atom MD to have the protein starting in the solution and "finding" its binding orientation on the membrane. Additionally, it takes longer for the protein to explore the set of surrounding protein:lipid contacts using all-atom MD simulations, and alternate protein states may be missed if they are not reached within the simulated time.

To address the slow diffusion rate of lipids, we use the Highly Mobile Membrane-Mimetic (HMMM) model of the membrane developed by the Tajkhorshid group.[28] This model system replaces a portion of the lipid tails with 1,2-dichloroethane (DCLE) (**Fig. 3.10**), which removes steric clash that decelerates lipid diffusion while maintaining the lipid features that form contacts with peripheral membrane proteins. The use of this model has previously allowed us to capture binding events of the murine TIM proteins within nanosecond timescales.[21] Within the work shown here, simulations of hTIM3:membrane systems were done for both docked and undocked starting states. For an undocked starting state, hTIM3 was placed 10-15 Å away from the membrane and a binding event was allowed to occur through the simulation. For a docked starting state, the protein was placed at varying orientations (further detailed in **Chapter 6**) at the membrane and the HMMM simulation was used to accelerate the search for a stably-bound orientation.

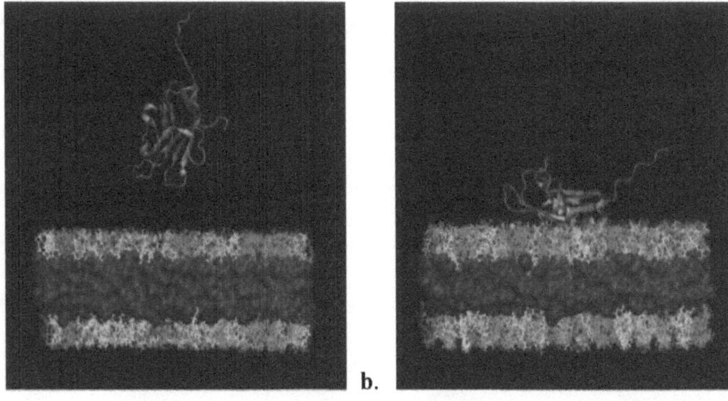

Fig. 3.10. Side-view of HMMM model for lipid bilayer with hTIM3 in solution. Protein (cyan) with calcium ion (yellow) in PS-binding pocket **a.** is placed ~10 Å away from the membrane for undocked simulations, and **b.** at the membrane for docked simulations. Short-tailed POPC and POPS are shown in pink and yellow, respectively. Surface representation of the internal portion of the membrane, DCLE, is shown in blue.

3.3.4.3. MD Simulations: Protocol

HMMM simulations of hTIM3 systems were set-up using the HMMM Builder[29] in the CHARMM-GUI.[21,30] For undocked simulations, a Ca^{2+}-bound protein was placed ~10-15 Å above a symmetrical 7:3 PC:PS bilayer membrane with approximate box sizes of $x = 100$ Å and $y = 100$ Å. For all simulations, the preset values of the CHARMM-GUI for Nanoscale Molecular Dynamics[31] (NAMD) input generation were used except for the following: the system was neutralized with 150 mM NaCl, WYF parameters[32] were used for cation-pi interactions under CHARMM36m force field, and equilibration was done under NP_nAT at 310 K. For all starting protein states, 5-10 systems were initialized to account for various distributions of lipids in the membrane. The generated systems were equilibrated with the default settings of the CHARMM-GUI HMMM Builder, and then simulated for up to 200 ns with 2 fs time steps. Simulations that were unlikely to result in a binding event (e.g., loss of the Ca^{2+} ion in the PS-binding pocket or protein drifted away from membrane) were terminated early (see **Chapter 6** for full discussion on stability of hTIM3 systems). A few systems were chosen to be extended beyond 200 ns to μs time scales. These were done by our collaborator Dr. Jeffrey Weber at IBM Watson Research Center.

Due to limitations of the CHARMM-GUI in initializing a HMMM membrane for a docked protein with a POPS placed in the binding pocket, docked simulations were initialized using a two-step process. First, the Bilayer Builder of the CHARMM-GUI was used to build a full membrane system[33], where hTIM3 with a Ca^{2+} ion and a full-length POPS already in the binding pocket was placed at the membrane. For this, the protein was translated along the z axis in order to align the headgroup of POPS with the headgroups of the other lipids in the membrane (**Chapter 6** further details the chosen POPS states and protein orientations). The same settings in the CHARMM-GUI were used as described above as for the HMMM Builder.

Next, the output *step5_assembly.str*, *step5_assembly.psf*, and *step5_assembly.crd* files of the Bilayer Builder were used in CHARMM-GUI's HMMM Builder, using the option to convert a full membrane system to a HMMM one. To account for the extra POPS lipid in the system, the *step5_assembly.str* file was modified to increase the number of lipids in the bottom leaflet by 1 (SET NLIPBOT = 151), and to include the following lines: SET POSID = POT; SET NEGID = CLA. Additionally, the *step5_assembly.psf* and *step5_assembly.crd* files were modified for the CHARMM-GUI to read the POPS that was placed into the binding pocket of hTIM3 as a part of the membrane. This was done by changing the segment id of the POPS to "MEMB" and its residue number to one more than the previous number of lipids in the membrane (i.e., if there were 310 lipids in the initialized membrane, the new residue number of this POPS would become 311). All other settings were chosen to be the same as for the HMMM Builder, and the output files of the

builder were run as described above.

3.4. Protein Purification

3.4.1. Introduction to Protein Purification

Protein purification within this book covers several protein systems, including the TIM proteins,

α-synuclein, and MFGE8. The section below details the final protocols used to purify each

protein, and also briefly discusses methods that were unsuccessful. All protein purification steps

were done in collaboration with Erin Adams at the University of Chicago.

3.4.2 TIM Family

Insect cell codon-optimized gene blocks for the IgV domains of human TIM1, 3, and 4 with C-

terminal 6x His tags were cloned into the pAcGP67a vector for baculovirus transfection (sequences

gene block and primers in **Tables 3.1** and **3.2**). Protein was then produced in Hi5 cells, as

described.[18] Protein productions used Lonza cell media (cells infected at a density of 1.5 million

cells/mL) or ESF cell media (cells infected at a density of 1.million cells/mL). Following a three-

day expression, cells were spun down at 8g and the supernatant was filtered.

Buffers were added to the supernatant for final concentrations of 1mM HEPES and 15mM NaCl

(1x HBS, pH 7.2), 1mM MgCl, 2mM imidazole, and .5mM $NiSO_4$, and the supernatant was

then stirred with Ni-NTA beads for four hours at 4 °C. Ni-NTA beads were then washed with 30mh

of Buffer B (1 mM HEPES, 35 mM NaCl, 2mM imidazole, pH 7.2), and the final protein was

eluted with 12mL of Buffer C (1 mM HEPES, 35 mM NaCl, 2mM imidazole, pH 7.2). 5mM

phenylmethylsulfonyl fluoride (PMSF) and Carboxypeptidase-A was then added for overnight,

room-temperature incubation for cleavage of the C-terminal His tag. Carboxypeptidase-A was then

inactivated with 5mM $CaCl_2$. At this step, Endo F was added to the hTIM3 sample for four hours at 37 °C for deglycoslyation. All three proteins were then concentrated and purified through size-exclusion chromatography (Superdex S200 or Superdex S75 column) in Buffer D (10 mM HEPES, 150 mM NaCl, 0.1% NaN$_3$, pH 7.2). Ca^{2+} and PS dependence for membrane binding for hTIM1 and hTIM4 were tested using the tryptophan fluorescence assay, while hTIM3 was tested using a sucrose sedimentation assay.

Several attempts were made to refold hTIM3 from a GST-hTIM3 construct following protocol from Gandhi et al.[34]. While we were able to express and purify the GST-hTIM3 inclusion body, we were unable to successfully refold the protein without precipitation in the subsequent purification and cleavage steps. While the GST construct would refold, as indicated by the protein binding to glutathione agarose beads, hTIM3 would aggregate following elution and cleavage, suggesting an improper fold. Refold attempts included varying the duration of the refold from 1-14 days, and the addition of Ca^{2+} and/or benzoic acid (**PDB 6DHB**) to the refold buffer in attempts to stabilize the IgV domain of hTIM3.

3.4.3. α-Synuclein

3.4.3.1. α-synuclein: Wild Type Purification

N-terminal acetylated WT α-synuclein and all associated mutants were expressed and purified as previously described.[35] The following protocol describes the standard protocol and includes a few adjustments that were made to purify ^{15}N and Alexa488-labeled protein. In short, the sequence for WT α-synuclein was encoded into a pET-21a backbone from which all mutants were made using site-directed mutagenesis. BL21 cells that were pretransformed with a pNatB plasmid that encodes for the N-α-acetyltransferase were transformed with the pET-21a plasmids. From a starter culture of 25 mL, the BL21 cells were grown at 37 °C under Ampicillin and Chloramphenicol resistance.

Protein expression was induced with 1 μM of isopropyl β-D-1-thiogalactopyranoside (IPTG) for 4 h at 37°C at a density corresponding to $OD_{600} = 0.6$.

The cells were then pelleted for 15 min at 7,000 g and resolubilized in Buffer E (20 mM Tris-HCl, pH 8.0, 5 mM EDTA, 1 mM PMSF, pinch of DNAase), after which the cells were lysed by sonication for 10 min, and then the pH was adjusted to 3.5 with NaOH. At this pH, other cellular proteins will precipitate, while α-synuclein remains soluble. Cell debris was then pelleted by centrifugation at 18,000 g for 15 min. The supernatant containing the N-acetylated α-synuclein was adjusted to pH 7.0 with HCl. Ammonium sulfate was added to 50% saturation, and the protein was rotated at 4 °C for 1 h to precipitate out α-synuclein. The final precipitate containing the mostly pure protein was isolated as a pellet through centrifugation at 18,000 g for 30 min.

The pellet was resuspended in 3 mL of water, and the protein was dialyzed overnight at 4 °C into DI water using 0.5–3 mL capacity, 3.5k MWCO Slide-A-Lyzer Dialysis Cassettes. Finally, the dialyzed protein was injected onto a Superdex 200 10/300 column and eluted at ~15 mL with HBS buffer (10 mM HEPES pH 7.2, 150 mM NaCl, 0.1% azide). Fractions containing the protein were pooled, aliquoted, and flash-frozen with liquid nitrogen, and stored at -80 °C until use. Tryptophan fluorescence was used to monitor protein samples for aggregation or misfolding.

Of note, α-synuclein is highly susceptible to aggregation and misfolding during dialysis. While the WT and most of the mutants were purified with no significant differences in yield, protein preparations with mutations at the A53 site (A53E, A53T, and A53V) were much more susceptible to misfolding. Despite appearing to be in monomer form following their elution off the sizing column, upon thawing the tryptophan fluorescence spectra would vary among protein preparations, occasionally appearing as an intense left-shifted peak, indicating tryptophan burial into a hydrophobic core. While it is not clear if this burial is aggregated or misfolded protein, the protein does not appear to bind to lipid and is therefore not in a comparable intrinsically disordered state. Thus, care should be taken with these mutants to ensure that the protein is unfolded prior to use as determined by tryptophan fluoresce emission spectra.

3.4.3.2. α-synuclein: ^{15}N-labeling

^{15}N-labeled α-synuclein was purified as described above with a few minor modifications. BL21 cells were grown to a density corresponding to $OD_{400} = 0.4$, at which the cells were pelleted at 7000 rpm for 15 min. The cells were then carefully resuspended in 1 L of minimal media composed of (1x M9 minimal media, 18.7mM $^{15}NH_4Cl$, 2mM $MgSO_4$, 0.1 mM $CaCl_2$, 0.4% glucose), using 5x M9 minimal media (239 mM Na_2HPO_4, 110 mM KH_2PO_4, 42.8 mM NaCl) and allowed to grow under Ampicillin and Chloramphenicol resistance for an hour at 37 °C. The suspension was then moved to room temperature and overnight protein expression was induced with 1 μM of isopropyl β-D-1-thiogalactopyranoside (IPTG). No other changes were made to the protocol, and ^{15}N-labeled α-synuclein behaved identically to non-labeled α-synuclein except for the lower protein yields.

3.4.3.3. α-synuclein: Alexa488-Fluorophore Labeling

For fluorescence polarization (FP) experiments, fluorophore-tagged α-synuclein was made from α-synuclein constructs that had a cysteine mutation. We initially chose the S9C, S42C, and G93C sites, as they are predicted to face away from the lipid membrane by the helical wheel of the two α-helices in α-synuclein, and are less likely to affect protein binding. These sites are all located within the lipid-binding region of α-synuclein and are therefore suitable for quantifying protein:membrane interactions through FP. The disulfide forms of the protein were purified through the same protocol as WT α-synuclein and the following protocol was used to tag the protein at the mutated sites with the fluorophore.

Protein was diluted to a final concentration of 50-100μM in HBS, pH 7.2. A 10x molar excess of tris(2-carboxyethyl)phosphine (TCEP) was added to the solution to reduce the disulfide bonds between the α-synuclein monomers. Then, a 10mM stock of the fluorophore in DMSO was added for a final ratio of 10:1 of dye:protein. The solution was rotated overnight at 4°C in the dark, after which it was purified through size-exclusion chromatography on a Superdex S200 column. Final protein concentrations and the degree of fluorophore labeling (DOL) were determined using **Eq. 3.23** and **3.24** below from the protein's absorption at wavelengths of 280 nm and 494 nm, the protein's molar extinction coefficient of $\varepsilon = 5120$ cm^{-1}M^{-1}, and the fluorophore's molar extinction coefficient of $\varepsilon = 71{,}000$ cm^{-1}M^{-1}. We used FP to determine if the fluorophore's locations interfered with α-synuclein's ability to bind lipid membranes, finding that the G93C site was optimal for labeling and quantification purposes. Meanwhile, the S9C and S42C sites saw a minor and large decrease, respectively, in protein binding, suggesting that the fluorophore location hindered the ability of the helices to insert into the membrane.

$$[Protein] = \frac{A_{280} - 0.11(A_{494})}{5120} \qquad \text{Eq. 3.23}$$

$$DOL = \frac{A_{494}}{71{,}000 * [Protein]} \qquad \text{Eq. 3.24}$$

3.4.4. Milk Fat Globule-epidermal Growth Factor 8

The C2 domain of *mus musculus* MFGE8 was purified through bacterial expression from a pET29b vector with a 6x His tag at the C-terminus.[6] From a 25 mL starter culture of BL21 (DE3) cells transformed with the vector, 1 L of cells under kanamycin resistance was shaken at 37 °C to an $OD_{600} = 0.6$. The flask was then transferred to 25 °C and protein expression was induced with 1 mM IPTG. After four hours, the cells were pelleted via centrifugation at 6,000 g for 10 min. Cell pellets were then resuspended at 4 °C at 10 mL/g in Buffer F (20 mM NaPO$_4$, 0.5 M NaCl, pH 7.3) with 1 mM PMSF and DNAase. The cells were then lysed by sonication for 10 min (3 sec on, 5 sec off) and the lysate was centrifuged at 22,000 rpm for 30 min. The supernatant was brought to 25 mM imidazole, pH 7.0 and incubated with Ni-NTA agarose beads for 2 hours at 4 °C. The Ni-NTA beads were then collected and washed with 3x of 10 mL of Buffer G (20 mM $NaPO_4$, 0.5 M NaCl, 25 mM imidazole, 1 mM PMSF, pH 6.5), and the protein was eluted with 10 mL of Buffer H (20 mM $NaPO_4$, 0.5 M NaCl, 200 mM imidazole, 1 mM PMSF, pH 6.5). To remove the imidazole, the protein sample was then desalted into Buffer I (20 mM $NaPO_4$, 0.5 M NaCl, pH 6.5) using Econo-Pac 10 DG Desalting Prepacked Gravity Flow Columns (Bio-Rad). Following desalting, the protein was concentrated to ~1 mL and purified through size exclusion chromatography with a S75 column. Protein purity and function were checked through gel electrophoresis and tryptophan fluorescence binding assays. Of note, MFGE8 is susceptible to aggregation and all purification steps should be done at 4 °C.

3.4.4.1. *MFGE8: Mutants*

Tryptophan fluorescence binding experiments of MFGE8 indicate that the protein may have multiple bound states across its binding curve with the possibility that multiple tryptophan residues undergo changes to their environment upon the addition of lipid. To explore which tryptophan emission spectra are altered by lipid binding, tryptophan to phenylalanine mutants (W26F, W33F, W55F, W65F, and W143F) were made for MFGE8 (primers for site-directed mutagenesis for these mutations are listed in **Table 3.2**). Of these, W26F and W33F were unstable and aggregated during the purification process, suggesting that they are essential to global protein structure. Tryptophan fluorescence measurements of the other mutants, W55F, W65F, and W143F, found no change in binding curve structure, suggesting that W55, W65, and W143 do not engage with the membrane.

3.4.5. Materials

3.4.5.1. *Materials: DNA Sequences for Cloning*

The following two tables include the DNA sequences of the gene blocks for codon-optimized expression of the TIM proteins (**Table 3.1**) and the primers for cloning and site-directed mutagenesis (**Table 3.2**). All DNA was ordered from Integrated DNA Technologies (Coralville, IA).

Table 3.1. Codon-optimized gene blocks for cloning into the pAcGP67a vector for insect cell expression of hTIM1, hTIM3, and hTIM4.

Protein	Codon-optimized gene block
hTIM1	GCCTTTGCGGCGGATCCTATGGTGAAGGTGGGTGGTGAAGCGGGTCCTTCCGTAACGCTTCCATGTCACTAT TCCGGCGCCGTTACCTCCATGTGTTGGAACCGTGGTTCATGTTCTCTCTTTACTTGTCAGAATGGCATCGTGT GGACCAACGGCACCCATGTCACATACAGGAAGGACACTCGTTACAAACTTCTCGGAGATTTGTCTAGGCG
hTIM3	GCCTTTGCGGCGGATCCTAGCGAAGTTGAATATCGTGCTGAAGTAGGTCAGAACGCTTACTTGCCCTGTTTC TACACTCCGGCCGCGCCTGGCAATCTCGTCCCAGTGTGTTGGGGTAAAGGCGCGTGTCCCGTATTCGAGTG CGGCAACGTGGTGCTCAGGACGGACGAACGCGATGTAAATTACTGGACTTCACGCTATTGGTTGAACGGCG
hTIM4	GCCTTTGCGGCGGATCCTAGCGAGACAGTCGTTACCGAAGTGCTTGGACATCGTGTGACTCTGCCATGTTTG TATTCTAGTTGGTCACATAATTCAAATAGTATGTGCTGGGGAAAGGATCAATGCCCATACTCTGGATGCAAGG AGGCGCTTATTAGAACAGATGGTATGCGTGTGACTTCCAGAAAAAGTGCCAAATACAGATTGCAAGGCAC

Table 3.2. Forward and reverse primers used for site-directed mutagenesis of each mutant (α-synuclein and MFGE8) and cloning of hTIM codon-optimized gene blocks into the pAcGP67a vector for insect cell expression.

Mutant	Forward Primer	Reverse Primer
αSyn, F4W	CATATGGATGTGTGGATGAAAGGTCTGAG	CTCAGACCTTTCATCCACACATCCATATG
αSyn, V15A	AAAGAAGGCGCGGTGGCTG	CAGCCACCGCGCCTTCTTT
αSyn, A30P	GAAGCGCCCGGCAAAACG	CGTTTTGCCGGGCGCTTC
αSyn, E46K	AGCAAAACCAAAAAAGGCGTGGTTC	GAACCACGCCTTTTTTGGTTTTGCT
αSyn, H50Q	GCGTGGTTCAAGGTGTGGC	GCCACACCTTGAACCACGC
αSyn, G51D	TGGTTCATGATGTGGCCACCG	CGGTGGCCACATCATGAACCA
αSyn, A53E	CATGGTGTGGAAACCGTTGCAGAAAAAACGAA	TCTGCAACGGTTTCCACACCATGAACCACGCC
αSyn, A53T	CATGGTGTGACCACCGTTGCA	TGCAACGGTGGTCACACCATG
αSyn, A53V	CATGGTGTGGTCACCGTTGCA	TGCAACGGTGACCACACCATG
αSyn, F94W	GGCAACCGGTGGCGTTAAAAAAG	CTTTTTTAACGCCACCGGTTGCC
αSyn, G93C	CGGCGGCAACCTGTTTCGTTAA	TTAACGAAACAGGTTGCCGCCG
hTIM1, for pAcGP67a	GCCTTTGCGGCGGATCCTATGGTGAAG	AAAAAAAAAAATCTAGAA GCTAATGATG
hTIM3, for pAcGP67a	GCCTTTGCGGCGGATCCTAGCGAAGTTG	TTCTAGAAGCTAATGATGATGGTGGTGA
hTIM4, for pAcGP67a	GCCTTTGCGGCGGATCCTAGCGAGACAG	AAAAAAAAAAATTCTAGAAGCTAG TGATGATGGTGGTG

Table 3.2, continued.

MFGE8, W26F	GCAGCTACAAGACATTTAACCTGCGTGCTT	AAGCACGCAGGTTAAATGTCTTGTAGCTGC
MFGE8, W33F	GTGCTTTTGGCTTTTACCCCCACTTG	CAAGTGGGGGTAAAAGCCAAAAGCAC
MFGE8, W55F	CAAGATCAATGCCTTTACGGCTCAGAGCAA	TTGCTCTGAGCCGTAAAGGCATTGATCTTG
MFGE8, W65F	AGTGCCAAGGAATTTCTGCAGGTTGACC	GGTCAACCTGCAGAAATTCCTTGGCACT
MFGE8, W143F	CTTCCAGTGTCCTTTCATAACCGCATC	GATGCGGTTATGAAAGGACACTGGAAG

3.4.5.2. Materials: Lipid Vesicle Preparation

Large unilamellar vesicles (LUVs) were used as the lipid source for all described α-synuclein and hTIM3 lipid-binding assays.[5] First, desired lipid concentrations at 12 or 18 mM were mixed from lipid stocks in chloroform (Avanti Polar Lipids, Alabaster, AL). The chloroform was first evaporated under nitrogen and then under vacuum overnight. The lipid sample was resuspended in HBS buffer, shaken at 40 °C for 40 minutes, and then taken through a freeze-thaw cycle five times. Lipids were then extruded using filters of the desired vesicle size. For all TIM experiments, vesicles were made using a hand-held extruder with filters with a diameter of 100 nm. For all α-synuclein experiments, vesicles were extruded using a Lipex extruder through filters with a diameter of 50 nm. Vesicle diameter and polydispersity for both extrusion methods were measured using dynamic light scattering (DLS, Zen3600 Malvern Nano Zetasizer) with final vesicle size measuring to be ~115 nm and ~70 nm, respectively. Final lipid concentration was measured using phosphate analysis.

3.5. References

1. Hokanson, D. E. & Ostap, E. M. Myo1c binds tightly and specifically to phosphatidylinositol 4,5-bisphosphate and inositol 1,4,5-trisphosphate. *Proceedings of the National Academy of Sciences* **103**, 3118–3123 (2006).

2.   Rossi, A. M. & Taylor, C. W. Analysis of protein-ligand interactions by fluorescence polarization. *Nat Protoc* **6**, 365–387 (2011).

3.   De Rosa, L., Di Stasi, R., Romanelli, A. & D'Andrea, L. D. Exploiting Protein N-Terminus for Site-Specific Bioconjugation. *Molecules* **26**, 3521 (2021).

4.   Kraft, C. A., Garrido, J. L., Leiva-Vega, L. & Romero, G. Quantitative Analysis of Protein-Lipid Interactions Using Tryptophan Fluorescence. *Sci Signal* **2**, (2009).

5.   Kerr, D. *et al.* How Tim proteins differentially exploit membrane features to attain robust target sensitivity. *Biophys J* **120**, 4891–4902 (2021).

6.   Suwatthee, T. *et al.* MFG-E8: a model of multiple binding modes associated with ps-binding proteins. *The European Physical Journal E* **46**, 114 (2023).

7.   Callis, P. R. [7] 1La and 1Lb transitions of tryptophan: Applications of theory and experimental observations to fluorescence of proteins. in 113–150 (1997). doi:10.1016/S0076-6879(97)78009-1.

8.   Ghisaidoobe, A. & Chung, S. Intrinsic Tryptophan Fluorescence in the Detection and Analysis of Proteins: A Focus on Förster Resonance Energy Transfer Techniques. *Int J Mol Sci* **15**, 22518–22538 (2014).

9.   Pfefferkorn, C. M., Jiang, Z. & Lee, J. C. Biophysics of α-synuclein membrane interactions. *Biochimica et Biophysica Acta (BBA) - Biomembranes* **1818**, 162–171 (2012).

10.   Greenfield, N. J. Circular Dichroism (CD) Analyses of Protein-Protein Interactions. in 239–265 (2015). doi:10.1007/978-1-4939-2425-7_15.

11.   Dodero, V. I. Biomolecular studies by circular dichroism. *Frontiers in Bioscience* **16**, 61 (2011).

12.   Hu, Y. *et al.* NMR-Based Methods for Protein Analysis. *Anal Chem* **93**, 1866–1879 (2021).

13.   Klein-Seetharaman, J. *et al.* Differential dynamics in the G protein-coupled receptor rhodopsin revealed by solution NMR. *Proceedings of the National Academy of Sciences* **101**, 3409–3413 (2004).

14.   Das, T., Acosta, D. & Eliezer, D. Interactions of IDPs with Membranes Using Dark-State Exchange NMR Spectroscopy. in 585–608 (2020). doi:10.1007/978-1-0716-0524-0_30.

15.   Schwarz, T. C. *et al.* High-resolution structural information of membrane-bound α-synuclein provides insight into the MoA of the anti-Parkinson drug UCB0599. *Proceedings of the National Academy of Sciences* **120**, (2023).

16.   Lee, J. H., Ying, J. & Bax, A. Nuclear Magnetic Resonance Observation of α-Synuclein Membrane Interaction by Monitoring the Acetylation Reactivity of Its Lysine Side Chains. *Biochemistry* **55**, 4949–4959 (2016).

17.   Scott, J. L., Musselman, C. A., Adu-Gyamfi, E., Kutateladze, T. G. & Stahelin, R. V. Emerging methodologies to investigate lipid–protein interactions. *Integrative Biology* **4**, 247 (2012).

18.   Kerr, D. *et al.* Sensitivity of peripheral membrane proteins to the membrane context: A case study of phosphatidylserine and the TIM proteins. *Biochimica et Biophysica Acta (BBA) - Biomembranes* **1860**, 2126–2133 (2018).

19.   Tiemeyer, S., Paulus, M. & Tolan, M. Effect of Surface Charge Distribution on the Adsorption Orientation of Proteins to Lipid Monolayers. *Langmuir* **26**, 14064–14067 (2010).

20.   Passmore, L. A. & Russo, C. J. Specimen Preparation for High-Resolution Cryo-EM. in 51–86 (2016). doi:10.1016/bs.mie.2016.04.011.

21.   Kerr, D. The Membrane Context of Phosphatidylserine Exposure Influences its Recognition by the TIM Proteins: A Structurally Guided Study. (University of Chicago, 2019).

22.   Sarkis, J. & Vié, V. Biomimetic Models to Investigate Membrane Biophysics Affecting Lipid–Protein Interaction. *Front Bioeng Biotechnol* **8**, (2020).

23.   Liu, W., Wang, Z., Fu, L., Leblanc, R. M. & Yan, E. C. Y. Lipid Compositions Modulate Fluidity and Stability of Bilayers: Characterization by Surface Pressure and Sum Frequency Generation Spectroscopy. *Langmuir* **29**, 15022–15031 (2013).

24.   Schlossman, M. *Liquid Surfaces and Interfaces.* (Cambridge University Press, 2012).

25.   Málková, Š. *et al.* X-Ray Reflectivity Studies of cPLA2α-C2 Domains Adsorbed onto Langmuir Monolayers of SOPC. *Biophys J* **89**, 1861–1873 (2005).

26.   Tietjen, G. T. *et al.* Molecular mechanism for differential recognition of membrane phosphatidylserine by the immune regulatory receptor Tim4. *Proceedings of the National Academy of Sciences* **111**, (2014).

27.   Chen, C.-H., Málková, Š., Cho, W. & Schlossman, M. L. Configuration of membrane-bound proteins by x-ray reflectivity. *J Appl Phys* **110**, (2011).

28.   Ohkubo, Y. Z., Pogorelov, T. V., Arcario, M. J., Christensen, G. A. & Tajkhorshid, E. Accelerating Membrane Insertion of Peripheral Proteins with a Novel Membrane Mimetic Model. *Biophys J* **102**, 2130–2139 (2012).

29.   Qi, Y. *et al.* CHARMM-GUI HMMM Builder for Membrane Simulations with the Highly Mobile Membrane-Mimetic Model. *Biophys J* **109**, 2012–2022 (2015).

30.   Jo, S., Kim, T., Iyer, V. G. & Im, W. CHARMM-GUI: A web-based graphical user interface for CHARMM. *J Comput Chem* **29**, 1859–1865 (2008).

31.   Brooks, B. R. *et al.* CHARMM: The biomolecular simulation program. *J Comput Chem* **30**, 1545–1614 (2009).

32. Khan, H. M., MacKerell, A. D. & Reuter, N. Cation-π Interactions between Methylated Ammonium Groups and Tryptophan in the CHARMM36 Additive Force Field. *J Chem Theory Comput* **15**, 7–12 (2019).

33. Wu, E. L. *et al.* CHARMM-GUI *Membrane Builder* toward realistic biological membrane simulations. *J Comput Chem* **35**, 1997–2004 (2014).

34. Gandhi, A. K. *et al.* High resolution X-ray and NMR structural study of human T-cell immunoglobulin and mucin domain containing protein-3. *Sci Rep* **8**, 17512 (2018).

35. Chung, P. J. *et al.* α-Synuclein Sterically Stabilizes Spherical Nanoparticle-Supported Lipid Bilayers. *ACS Appl Bio Mater* **2**, 1413–1419 (2019).

CHAPTER 4

PD-ASSOCIATED MUTATIONS ALTER EQUILIBRIUM BOUND STATE OF α-SYNUCLEIN

4.1. Introduction

The progression of Parkinsons's Disease (PD) does not require a genetic predisposition, and in many cases, patients develop PD without genetic markers within the SNCA gene.[1] However, the increased severity and earlier onset in patients with PD-associated mutations point to α-synuclein's role within the disease, which is further accentuated by the implication of α-synuclein:lipid membrane interactions within the common markers of PD.[2] In the field's attempt to clarify α-synuclein's pathophysiological role, these PD-associated mutants provide an important exploratory avenue that could provide insight into dysfunctional states of WT α-synuclein. For example, conserved features of the lipid-binding properties of these mutants may point to specific states or behaviors of α-synuclein that can initiate downstream aggregation.

However, the number of parameters and protein preferences within the α-synuclein:lipid system obfuscates analysis and comparison of binding data across studies and PD-associated mutants. Within the system, a "lipid-bound state" can take on a variety of meanings, such as partial vs. full engagement with the membrane, and these states are heavily influenced by the sample and sample environment, including temperature[3] and the presence of certain ions (Ca^{2+}),[4] along with sample preparation. Furthermore, the preference of α-synuclein for specific lipid compositions along with its curvature sensitivity[5] requires near identical sample conditions for cross-study comparison, and to date we are unaware of any studies that have examined α-synuclein:lipid binding properties under the same conditions[6–8] across all eight PD-associated point mutations: V15A, A30P, E46K, H50Q, G51D, A53E, A53T, and A53V.

The following chapter demonstrates how characterizing the lipid-binding behaviors of WT α-synuclein and PD-associated mutants through a single, uniform approach can reveal subtle detail that may be otherwise hidden by the complexities of the system. Within this work, we use a tryptophan fluorescence assay to probe the effect of PD-associated mutations on the bound states of the two α-helices in α-synuclein. In addition to demonstrating the utility of this technique in studying α-synuclein:membrane binding, our work uncovers a previously-unknown conserved feature across the PD-associated mutants.

4.2. Lipid Depletion Model

In analyzing the α-synuclein:lipid system, we must first establish a suitable quantitative analysis that represents the entirety of the binding interactions. While the affinities for many protein:ligand systems can be adequately analyzed through the Michaelis-Menten binding equation,[9] this general model cannot be directly applied to quantify α-synuclein:lipid interactions. The extended nature of its two helices requires α-synuclein to interact with a large surface area of the membrane as the protein's binding behavior is primarily mediated by protein:lipid contacts across the full length of the helices. While individual lipid species, such as phosphoserine (PS), may increase the affinity of α-synuclein for the membrane,[10] the true "binding site" of α-synuclein forms from a collection of membrane parameters. Within this context, a "binding site" can be defined as any collection of lipid species that accommodates all the lipid contacts that α-synuclein requires across the length of both α-helices in their association with the membrane. Previous research within the field has highlighted several key membrane parameters that affect α-synuclein's affinity for lipid membranes, including vesicle curvature and lipid tail/headgroup composition.[5,10] However, no single parameter can solely characterize the binding of α-synuclein, as multiple configurations of lipid species can provide the necessary protein:lipid contacts within the membrane. Therefore, the configuration of lipid species that comprise a "binding site" may be highly varied, resulting in a diversity of distinct lipid configurations that constitute "binding sites" for α-synuclein. As such, this must be accounted for when interpreting the affinity of α-synuclein for its "ligand."

α-synuclein directly interacts with a few dozen lipids but requires many more lipids to indirectly support those interactions. For α-synuclein to fully insert into the membrane, lipids must assemble or rearrange to locally contain the necessary lipid contacts and available surface space under the protein footprint. The local concomitance of these necessary parameters occurs with a much lower density on the membrane surface than the density of any one of the parameters. As a result, the density of a binding site is much lower on the vesicle surface than the density of any individual lipid species that comprises that binding site.

When the measured bound fraction of protein is dictated by the availability of the limited binding sites rather than the affinity of the protein for these sites, the protein is found to be within the "lipid depletion regime." Within this regime (i.e., high protein concentration with respect to lipid binding sites), all available binding sites are occupied by proteins. Many *in vitro* binding assays with high protein concentrations, such as NMR and CD, are confined to the lipid depletion regime, and this regime may also be physiologically relevant, as local concentrations of α-synuclein within the synapse can reach ~22 μM,[11] and the surface area of synaptic vesicles has a high density (60% occupancy by mass [12]) of other proteins. Both the complexity of the identity for a "binding site" and the protein concentration dependence of the lipid depletion regime render quantitative analysis of α-synuclein's interactions with lipid membranes highly dependent on experimental conditions. Within the availability-dictated binding of the lipid depletion regime, solely relying on the traditional dissociation constant (K_d) as a complete descriptor of binding can complicate our understanding of the role of α-synuclein:membrane interactions in PD progression, and other binding parameters should be used in parallel to interpret the multidimensional interactions between the protein and lipid species.

Within the lipid depletion regime, the binding site density (σ) can be used in parallel with the K_d to better characterize the binding behavior of the protein at low binding-site availability and to serve as a more stable readout of protein:membrane interactions.[13,14] Briefly, the binding site density represents the lipid density at which the membrane parameters that are required for a single protein to bind occur (further detailed in **Section 4.2.1**). For example, a σ value of 0.01 would indicate that on average a binding site that accommodates all the required protein:lipid contacts is available ~100 lipid molecules across both leaflets of the lipid bilayer.

Within our system, σ can be used to further our understanding of how mutations within α-synuclein affect the protein's interactions with lipid membranes by tracking how changes to the protein:lipid system alter the density at which the required contacts occur. The mutations themselves may also alter the identity of the binding sites, as changes to the structure of the protein may require a separate collection of membrane parameters to accommodate its insertion. For example, the E46K mutation within α-synuclein may alter its lipid charge preferences, while the A53V mutation may alter preferences for membrane fluidity or curvature. While the binding site density does not provide lipid-specific information, it does probe how much the global lipid environment is able to accommodate a given protein in terms of the rarity of binding sites. This allows for a comparison of the behavior of various proteins across identical membrane conditions.

The binding of α-synuclein is often treated as binary, but the interactions of α-synuclein with the membrane comprise a collection of states in which the two helices formed upon binding to the membrane are engaged with the membrane to varying extents. Within the lipid depletion regime, the binding site density of each helix's bound state is critical as it dictates the distribution of these bound states on the membrane surface. Therefore, quantifying interactions between α-synuclein and the lipid membrane within the depletion regime requires helix-specific resolution. By comparing **σ** values from the binding behavior of various regions of the protein, we gain insight into the equilibrium binding of the protein within the depletion regime.

4.2.1. Binding Site Density

The explicit, mathematical dependence of the bound fraction of protein (*f_bound*) on the **σ** and **K_d** parameters can be quantified through the following depletion model,[14] where **[P]_tot** and **[L]_tot** are the total protein and lipid concentrations, respectively (**Eq. 4.1**).

$$f_{bound} = \frac{1}{2[P]_{tot}}\left(\sigma * K_d + [P]_{tot} + \sigma * [L]_{tot} - \sqrt{(\sigma * K_d + [P]_{tot} + \sigma * [L]_{tot})^2 - 4[P]_{tot}\sigma * [L]_{tot}}\right) \quad \textbf{Eq. 4.1}$$

When the protein concentration is greater than the combined contributions to the affinity from **σ** and **K_d** (**[P]_tot** > **σ*K_d**), the fraction bound can be rewritten as **Eq. 4.2** and then approximated by **Eq. 4.3** using (**σ*K_d/[P]_tot**) = 0 for lower lipid concentrations, which in turn reduces to **Eq. 4.4**.

$$f_{bound} = \frac{1}{2} + \frac{\sigma * K_d}{2[P]_{tot}} + \frac{\sigma * [L]_{tot}}{2P_{tot}} - \frac{1}{2}\sqrt{\left(\frac{\sigma * K_d}{P_{tot}} + 1 + \sigma * \frac{[L]_{tot}}{P_{tot}}\right)^2 - 4\sigma * \frac{1}{P_{tot}} * [L]_{tot}} \quad \textbf{Eq. 4.2}$$

$$f_{bound} \cong \frac{1}{2} + \frac{\sigma * [L]_{tot}}{2P_{tot}} - \frac{1}{2}\sqrt{\left(1 + \sigma * \frac{[L]_{tot}}{P_{tot}}\right)^2 - 4\sigma * \frac{1}{P_{tot}} * [L]_{tot}} \quad \textbf{Eq. 4.3}$$

$$f_{bound} \cong \frac{\sigma * [L]_{tot}}{P_{tot}} \quad \textbf{Eq. 4.4}$$

Within this region, the binding is within the depletion regime, where the binding curve at lower lipid concentrations can be approximated by a line with a slope proportional to **σ** (**Fig. 4.1**). As shown, even a $[P]_{tot}/\sigma*K_d$ ratio as low as ~3 (green curve in **Fig. 4.1**) results in the binding curve that can be well-approximated at lower lipid concentrations by a line with a slope equal to $\sigma/[P]_{tot}$.

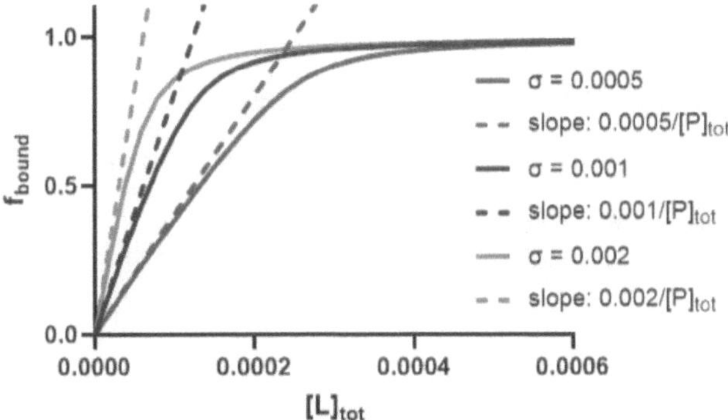

Fig. 4.1: Modeled binding curves (solid lines) for **Eq. 4.1** at varying binding site densities σ = 0.0005 (red), σ = 0.001 (blue), and σ = 0.002 (green) at constant $[P]_{tot}$ = 125 nM, and K_d = 20 µM. Dashed lines corresponding to a slope of $\sigma/[P]_{tot}$ for each curve are shown to demonstrate the region corresponding to the lipid depletion regime at lower lipid concentrations.

While we fit our α-synuclein:lipid binding curves using both **σ** and **K_d** parameters, the **σ** parameter holds critical information about the binding behavior of the protein under limited lipid availability, as the binding site density **σ** represents the lipid density at which the necessary lipid membrane parameters assemble to form a binding site for the protein. These lipid membrane parameters are reflective of protein contacts with lipids within the entire membrane environment. While these membrane parameters differ across various membrane compositions, analysis of **σ** values between mutant proteins on the same membrane allows for direct comparison of membrane parameters needed to accommodate protein changes. Therefore, comparison of the **σ** parameter between the two helices allows us to examine how binding site availability differs between the two helices for both wild-type (WT) and mutant α-synucleins.

4.3. α-Synuclein: Tryptophan Fluorescence Spectroscopy: WT

We first characterized the binding of the two helices of WT α-synuclein to LUVs with a lipid composition of 55:20:15:10 DOPC:DOPS:DOPE:CHOL using tryptophan fluorescence spectroscopy. To independently probe at each helix, we used both the Helix 1 reporter (F4W mutant) and the Helix 2 reporter (F94W mutant) (**Fig. 4.2**). The binding curves for the two reporters were each fit to the lipid depletion model[14] (**Eq. 4.1**) with fit parameters of the dissociation constant, K_d, and the binding site density, σ.

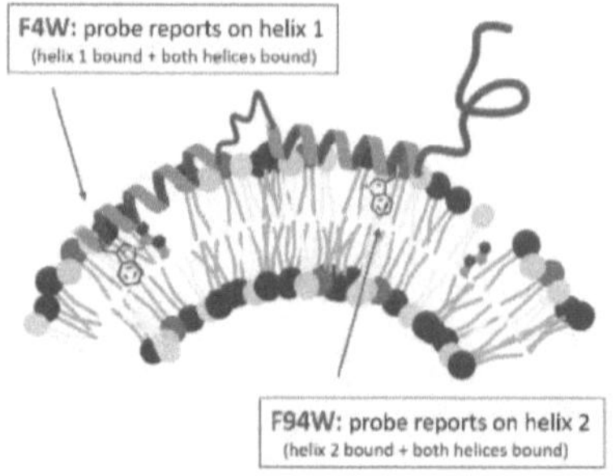

Fig. 4.2: Location of Helix 1 Reporter (F4W) and Helix 2 Reporter (F94W). Tryptophan side chain (red) inserts into the lipid membrane.

The binding curves for Helix 1 (H1) Reporter and Helix 2 (H2) Reporter are shown in **Fig. 4.3a** and **4.3b**, respectively, and their corresponding site density σ values, obtained from fitting the curves to **Eq. 4.1**, were found to be $\sigma_{H1} = 0.00113 \pm 0.00006$ and $\sigma_{H2} = 0.00110 \pm 0.00021$ in units of inverse lipids. Converting to units of lipids, we see that a binding site that accommodates a single helix occurs in every ~909 lipids across the bilayer. Furthermore, fitting resulted in dissociation constants for Helix 1 and Helix 2 reporters of $K_{d,\,H1} = 23.9 \pm 3.62$ μM and $K_{d,\,H2} = 16.5 \pm 8.49$ μM, respectively. Within error of one another, this suggests that the helices have approximately similar affinities for the lipids directly involved in the binding site. Overall, binding curves for the Helix 1 Reporter and Helix 2 Reporter alone suggest that the two helices of WT α-synuclein equally engage with the membrane.

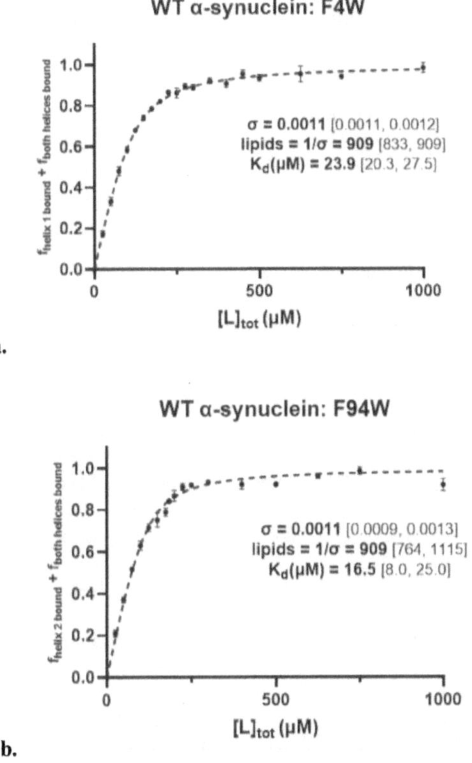

Fig. 4.3. Binding curve of WT α-synuclein.

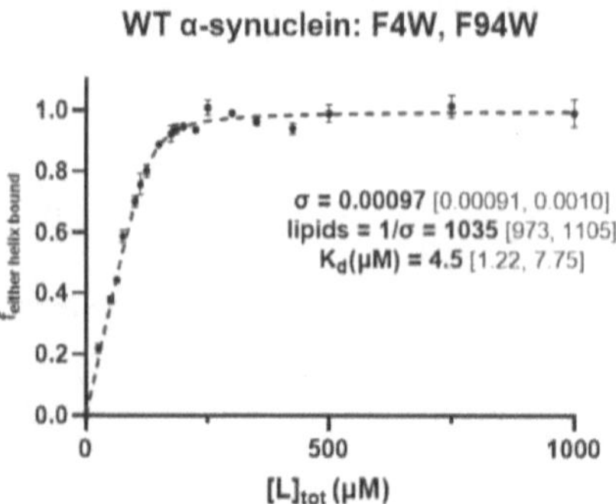

c.

Fig. 4.3, continued. Binding curve of WT α-synuclein at 125 nM protein concentration, using **a.** Helix 1 Reporter, **b.** Helix 2 Reporter, and **c.** Dual Reporter. The best fit obtained using **Eq. 4.1** is shown as a dashed line for each curve. Parameter fits are shown with a 95% confidence interval.

We next performed a similar binding experiment with WT α-synuclein with the Dual Reporter (DR), where σ_{DR} and $K_{d, DR}$ were fit to 0.0009 ± 0.000062 1/lipids and 4.5 ± 3.3 μM, respectively (**Fig. 4.3c**). The similar σ values of the single vs. dual reporter systems indicate that the two helices of WT α-synuclein seek out similar membrane parameters in binding, suggesting that the bound state of each helix is dependent on the binding parameter preferences of the other helix. Altogether, these data show that the equilibrium bound state of WT α-synuclein leans towards having both helices engaged with the membrane within the lipid depletion regime (**Fig. 4.4**).

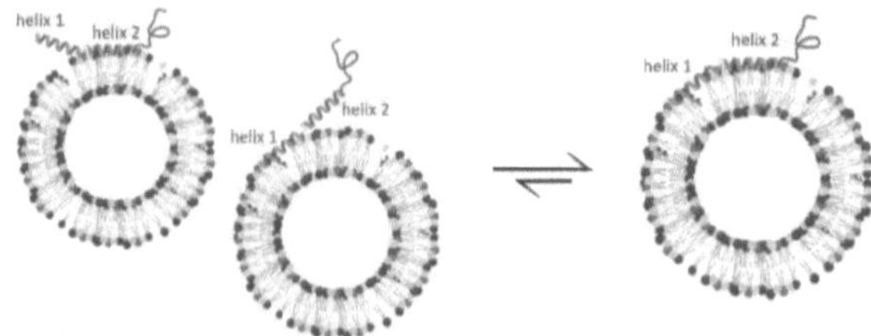

Fig. 4.4. Cartoon depiction of the equilibrium of WT α-synuclein within the depletion regime.

4.4. α-Synuclein: Tryptophan Fluorescence Spectroscopy of the E46K PD-Associated Mutant

This reporter system formed by the trio described above was used to examine how PD-associated mutants differentially affect the binding of the two helices of α-synuclein. As a case study, we generated α-synuclein with the PD-associated mutation E46K containing either the singular Helix 1 or Helix 2 Reporter and performed the tryptophan fluorescence assay to generate binding curves. The fitted σ values for Helix 1 Reporter (E46K) and Helix 2 Reporter (E46K) were σ_{H1} = 0.00083 ± 0.00006 and σ_{H2} = 0.0018 ± 0.00021 in units of inverse lipids, reflecting a binding site for the helices every ~1,198 and ~556 lipids across the bilayer, respectively (**Fig. 4.5a,b**). We can further compare WT and E46K results by examining the linear region of the bound fraction for each Reporter (**Fig. 4.5c**). The data indicates that a binding site accommodating Helix 2 of E46K is more abundant on the membrane than for Helix 1—a difference that does not exist for WT α-synuclein. While previous studies have suggested that the E46K mutation increases the affinity of the entire α-synuclein for lipid membranes,[15,16] our results here indicate that this may be a property largely conferred onto the protein by increased binding site availability for Helix 2 on the membrane.

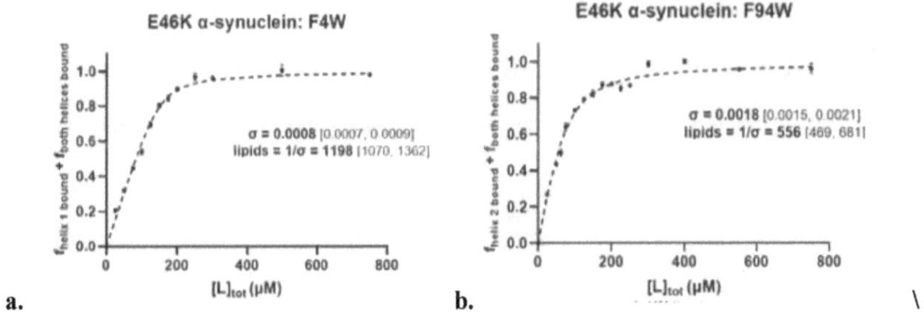

Fig. 4.5. a. Binding curve of E46K α-synuclein at 125 nM protein, using Helix 1 Reporter (left), and Helix 2 Reporter (right). Best fit curves using **Eq. 4.1** are shown as dashed lines. Parameter fits are shown with a 95% confidence interval.

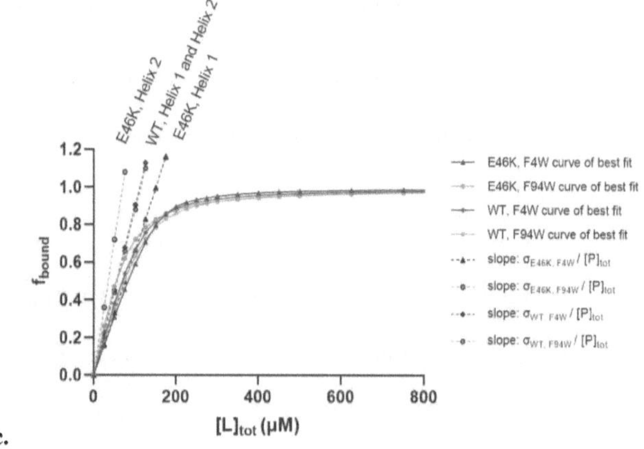

Fig. 4.5, continued. c. Best fit curves of WT and E46K α-synuclein for Helix 1 and Helix 2 Reporters (solid lines). Dashed lines corresponding to a slope of $\sigma/[P]_{tot}$ for each of the curves are shown to highlight the difference in binding site density in the E46K mutant in comparison to the WT.

Altogether, our findings from the reporter systems suggest that the E46K mutation within α-synuclein shifts the equilibrium towards Helix 2 within the lipid depletion regime (**Fig. 4.6**), and the lipid membrane contains fewer binding sites that accommodate the insertion of Helix 1 in comparison to Helix 2. These results are further supported by the binding curve for E46K α-synuclein with the Dual Reporter, where the slope of the binding curve of E46K in the depletion regime is closer to that of the Helix 1 Reporter (see first row in **Fig. 4.8**), demonstrating the dependence of the binding of the entire protein on the helix with lower binding site availability.

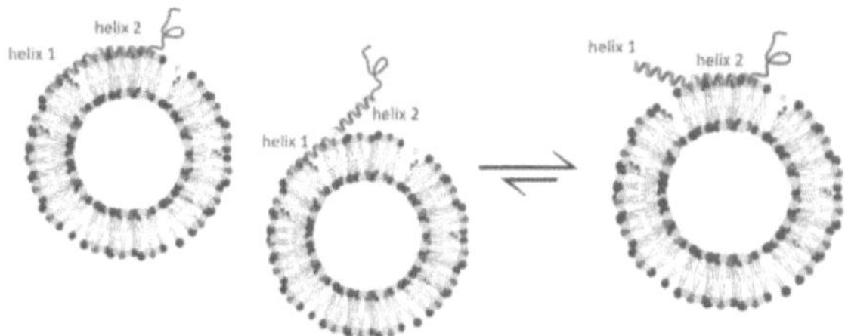

Fig. 4.6. Cartoon depiction of equilibrium in bound states of E46K α-synuclein. Of note, in comparison to WT, a state in which only Helix 2 is bound in a more dominant state for E46K than for WT α-synuclein.

4.5. α-Synuclein: Tryptophan Fluorescence Spectroscopy of PD-associated Mutants

A similar analysis was performed on six additional PD-associated mutants: V15A, A30P, H50Q, G51D, A53T, and A53V. The fit parameters from the binding curves for the three reporter systems are summarized in **Table 4.1**, while **Fig. 4.7** displays the number of lipids that constitutes a binding site for WT and all PD-associated mutants examined (**1/σ**). Across all seven PD-associated mutants, we see that a higher binding site density (lower **1/σ**) for Helix 2 than for Helix 1 alone, or for both helices (**Fig. 4.8**).

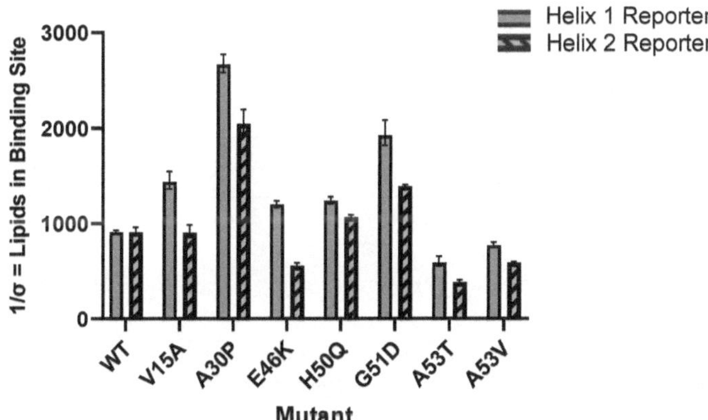

Fig. 4.7. Inverse of the binding site density (**1/σ**) for WT and PD mutants of α-synuclein

Fig. 4.7, continued. The inverse of the binding site density ($1/\sigma$), which gives the number of lipids per binding site, is shown for the WT and each of the seven PD-associated mutants of α-synuclein. Data for each sample were fit to **Eq. 4.1** independently for the Helix 1 and Helix 2. Of note, parameter fits for all PD-associated mutants show a statistically significant difference between the Helix 1 and Helix 2 reporter ($p < 0.05$), see **Table 4.2**.

We performed an F-test on the WT and each PD-associated mutant of α-synuclein comparing each fit to that in which σ is constrained to be the same for both helix reporters. While WT α-synuclein exhibits no difference in the fit between the Helix 1 and Helix 2 Reporter binding curves, the PD-associated mutants all showed a statistically significant ($p < 0.05$) difference between σ fits for the two reporters, with σ for the Helix 1 Reporter being consistently lower (**Table 4.2**). Altogether, these results indicate that the helices of α-synuclein are not equally sensitive to PD-associated mutations, resulting in a shift in the equilibrium bound state of the protein towards the second helix.

Table 4.1. $1/\sigma$ values for the WT and PD-associated α-synuclein mutants. Binding curves from each reporter were fit to **Eq. 4.1**. Values in table are displayed with 95% confidence intervals.

Mutant	lipids = $1/\sigma$ from Helix 1 Reporter Fit	lipids = $1/\sigma$ from Helix 2 Reporter Fit
WT	909 [833, 909]	909 [764, 1115]
V15A	1433 [1158, 1879]	903 [714, 1227]
A30P	2663 [2341, 3086]	2045 [1671, 2635]
E46K	1198 [1070, 1362]	556 [470, 681]
H50Q	1236 [1107, 1399]	1066 [991, 1154]
G51D	1923 [1526, 2551]	1388 [1323, 1458]
A53T	588 [455, 862]	384 [309, 488]
A53V	769 [699, 909]	588 [543, 633]

Table 4.2: F-test results for each Reporter pair across WT and each PD-associated mutant.

Mutant	F-test Statistic for Helix 1 vs. Helix 2 Reporter	P value for Helix 1 vs. Helix 2 Reporter	Same σ between Helix 1 and Helix 2 Reporter?
WT	0.0002935	0.9864	Yes
V15A	9.51	0.0052	No
A30P	6.101	0.0214	No

Table 4.2, continued.

E46K	50.08	<0.0001	No
H50Q	6.053	0.0218	No
G51D	8.117	0.0078	No
A53T	5.023	0.0338	No
A53V	26.55	<0.0001	No

4.6. α-Synuclein: Tryptophan Fluorescence Spectroscopy of the Dual Reporter

The Dual Reporter system can corroborate results from the single reporter systems through a qualitative comparison between the binding curves from the Dual Reporter and the Helix 1 Reporter. As the Dual Reporter reports on three bound states of the protein (Helix 1 only, Helix 2 only, or both), its fully bound state would be one in which both tryptophan residues are inserted into the membrane, and the binding curve thus corresponds to a fractional distribution of the three bound states. With the fully bound state of the Dual Reporter requiring both Helix 1 and Helix 2 to be bound, and given that the membrane contains fewer binding sites that accommodate Helix 1 compared to Helix 2 for the PD-associated mutants, the binding curve for the Dual Reporter is expected to exhibit a slope at lower lipid concentrations that is closer to the one in the binding curve of the Helix 1 Reporter than that of the Helix 2 Reporter.

It is worth noting that bound fractions obtained from the Dual Reporter system cannot be cleanly compared to those from the Helix 1 or Helix 2 Reporters due to changes in the distribution of bound states at different lipid concentrations for the former. For single-reporter systems, the calculated bound fractions are directly dependent on the inserted proportion of the single-site tryptophan as the reporter only reads two states: inserted and non-inserted. However, for a dual-reporter system with the tryptophan emission spectrum composed of four states – fully unbound, Helix 1 bound, Helix 2 bound, and both helices bound, changes to the ratio of inserted reporters at Helix 1 vs. Helix 2 across the binding curve will result in shifts in the combined Dual Reporter spectrum within regions of changing distribution of bound states, with the portion of the binding

curve where the protein:lipid ratio exits the lipid depletion regime being affected most. Whereas

for WT α-synuclein, the bound ratio of the two reporters does not appear to change along the curve

(as indicated by the binding curves of the Helix 1 and Helix 2 Reporters), this is not true for the

PD-associated mutants.

While full quantitative analysis of the σ and $\mathbf{K_d}$ parameters is not possible from the binding curve

of the Dual Reporter system, the curve still allows for qualitative comparison of the Dual Reporter

to the two single reporters through an analysis of the slope at lower lipid concentrations. As the

slope is dependent on σ and $\mathbf{K_d}$ as shown in **Eq. 4.4**, the slope can be used as an indicator for the

overall binding of the protein for the membrane. For our analysis of the Dual Reporter system, we

fit through the linear region of each binding curve (**Fig. 4.8**) to yield an estimate of the slope of

the depletion regime. The slope for the Dual Reporter is found to be similar to that for the Helix 1

Reporter, supporting that the Dual Reporter system reads out on membrane-binding of the full

protein, and the binding plateau can only be reached upon the binding of Helix 1 which has a lower

binding site availability.

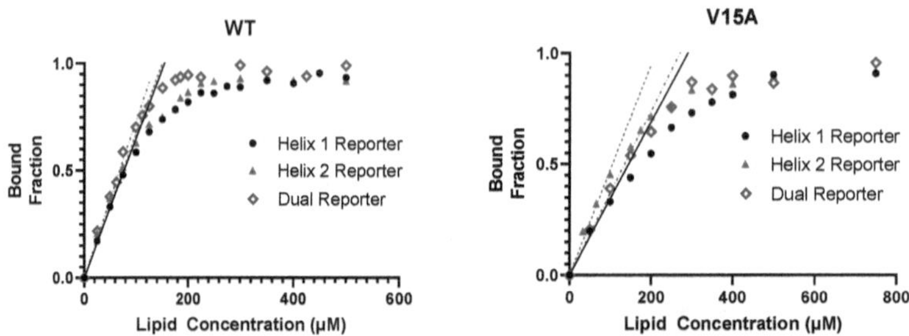

Fig. 4.8. Binding curves for WT α-synuclein and seven PD-associated mutants.

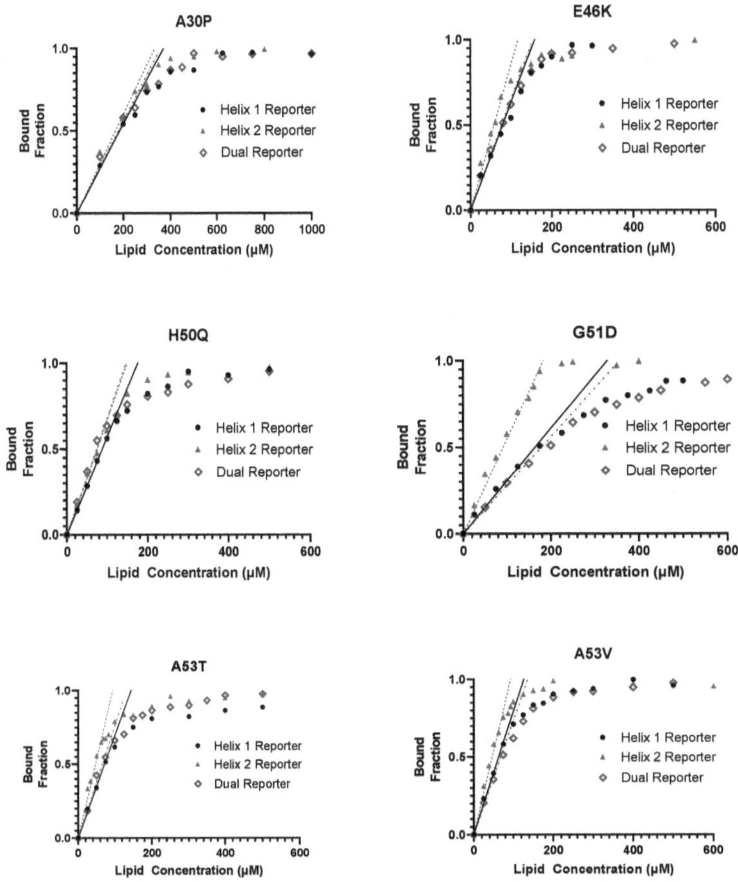

Fig. 4.8, continued. Binding curves for WT α-synuclein and seven PD-associated mutants. Bound fractions of full curves are shown for the Helix 1 Reporter (black), Helix 2 Reporter (red), and Dual Reporter (blue). Guide lines corresponding to the linear region of each binding curve are shown for each reporter. Of note, the lipid concentration axis is truncated from the full range examined to better show details at lower lipid concentrations, but each curve extends beyond the lipid concentrations shown to reach the binding plateau.

4.7. α-Synuclein: Tryptophan Fluorescence Spectroscopy of the A53E PD-Associated

Mutant

The described procedure and analyses were also performed for the A53E point mutation in α-synuclein, another identified PD-associated mutant. However, this point mutation was especially susceptible to misfolding or aggregation during protein purification and thawing. Despite several attempts, we were unable to use the protein for the binding experiments due to the presence of states other than its unfolded monomer form as indicated by a large left peak in the unbound protein tryptophan fluorescence spectrum of the Helix 1 Reporter (**Fig. 4.9a**) and a smaller left peak in the unbound spectrum of the Helix 2 Reporter (**Fig. 4.9b**) that were not present for the other mutants.

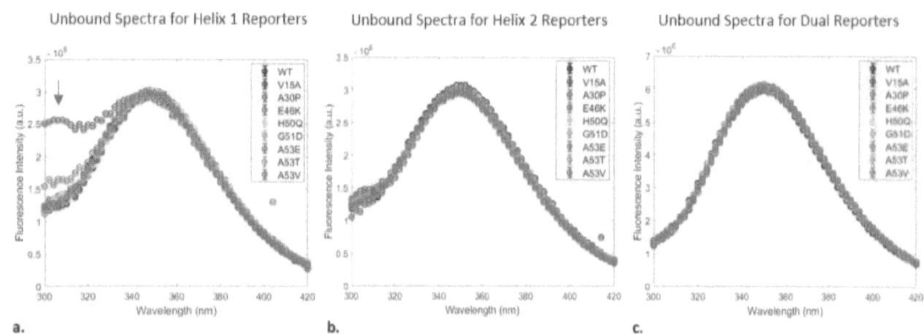

Fig. 4.9. Unbound tryptophan fluorescence emission spectra of **a.** Helix 1 Reporter, **b.** Helix 2 Reporter, and **c.** Dual Reporter for WT and all PD-associated mutants. Left peak in **a.** corresponds to a misfolded/aggregated state for A53E with the Helix 1 Reporter, as indicated by red arrow.

We speculate that this left peak in the unbound emission spectrum may be from protein that is misfolded or in an aggregated state. Upon addition of lipids, this left peak does not exhibit any shift in the fluorescence spectrum (**Fig. 4.10**), suggesting that this state of the protein does not associate with lipid membranes in a comparable manner to the unfolded monomer state.

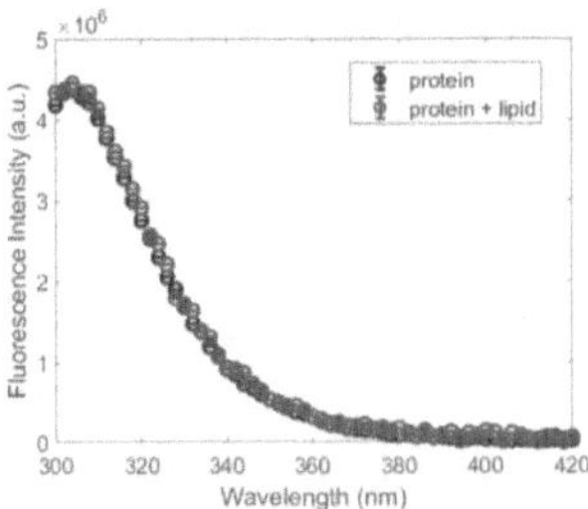

Fig. 4.10. Representative tryptophan emission spectra of misfolded or aggregated α-synuclein without (black) and with (purple) lipid.

Given that a portion of the protein in the experiment does not appear to engage in membrane binding, we surmise that the effective protein concentration is lower though we cannot provide a meaningful estimate of this concentration to support data analysis for the reporter systems, challenging our ability to carry out a quantitative comparison of the binding site availability for Helix 1 vs. Helix 2 as we did for the WT and other mutants. Of interest, tryptophan fluorescence may serve as a useful technique for future studies in monitoring the folded state or aggregation of α-synuclein, where site-specific burial of the tryptophan into a hydrophobic core would correspond to an increase in the left shoulder of the spectrum.

4.8. α-Synuclein: Conclusions for PD-associated Mutants

While PD-associated mutations are not required for the development and progression of Parkinson's Disease, these mutations do accelerate the onset and severity of associated symptoms.[2] In studying PD-associated mutants, we can gain insights into the potential neurotoxicity of WT α-synuclein. The true bound state of α-synuclein comprises a distribution of states in which the two helices of α-synuclein are bound to the membrane to varying extents; identifying which states are related to the dysfunction of α-synuclein is critical to uncovering the role α-synuclein binding to membranes plays within PD. In comparison to WT α-synuclein, a unifying feature of these PD-associated mutations emerges: across all seven mutants studied, there is a shift in the equilibrium bound states of α-synuclein towards the second helix. Therefore, the state in which the binding of the first helix of α-synuclein is destabilized and that of the second enhanced may play a central role in α-synuclein dysfunction within PD progression.

The tryptophan fluorescence assay used serves as a powerful tool to probe site-specific interactions between the α-helices of the protein and the lipid membrane. It allows us to quantitively characterize the effects of the PD-associated mutants on the bound state of each of the helices. Furthermore, the lipid depletion model employed here relates the binding of the protein to binding site availability, allowing for the binding site density, σ, to 1) reflect relevant membrane parameters and solution conditions for binding, and 2) directly gauge the relative propensity of the two helices for membrane binding across the PD-associated mutants. Of note, the specific locations of the Helix 1 Reporter and Helix 2 Reporter allow this assay to only probe for bound states in which the front and back ends of Helix 1 and Helix 2, respectively, are engaged with the membranes. Any partially bound state in which these residues are not inserted into the membrane will therefore be considered an unbound state within the analysis.

In comparison to a binary association model, the in-depth analysis of the two α-helices through an equilibrium model between states in which a single helix or both helices are bound reveals important features of PD-associated mutations. Our results indicate that all PD-associated mutations shift the equilibrium towards a bound Helix 2 state due to higher binding site availability for Helix 2 over Helix 1. Importantly, our results highlight how an appropriate binding model for quantifying protein:membrane interactions can reveal critical system information that may otherwise be obscured. Furthermore, within quantifying protein:lipid binding, it is important to consider the entire membrane context alongside specific membrane parameters.

The findings of this study may address some of the inconsistent results within the α-synuclein mutant literature. For example, some previous studies have reported increased lipid-binding for the E46K mutant,[15] while others[16] have shown negligible attenuation of binding. Our results here (increased binding site availability for the second helix, along with a mild decrease for the first) suggest that the method used in quantifying α-synuclein binding could significantly influence the binding data obtained. In the case of the E46K mutant, a method that reports on any bound state of α-synuclein (e.g., fluorescence microscopy[15]) would see an overall increase in lipid-binding, while methods that more directly probe the bound state of the helices (e.g., nuclear magnetic resonance[17] or circular dichroism) may measure little change[16] or a slight decrease in membrane association of the E46K mutant compared to WT. Our work thus underscores the importance of considering the distribution of bound states of α-synuclein in uncovering the effects of mutations and membrane parameters on the binding behavior of α-synuclein and its role within PD. It is worth noting that the higher protein concentrations and crowded lipid environment in the neuron may result in faster lipid depletion, and the effects of these mutations on the distribution of bound states of α-synuclein may be even more pronounced than those found in this study.

The α-synuclein field has established that the initial residues of the N-terminus are essential for triggering the α-helix formation and the subsequent binding of the protein.[18] Meanwhile, residues 61-95, termed the NAC region, have been proposed to have reduced lipid association and increased propensity to aggregate. Our results here point to a possibly larger role played by the second α-helix in stabilizing α-synuclein at the membrane than previously thought. As the position of the Helix 2 Reporter is at residue 94, any measured bound state for the second helix would require the engagement of its end with the membrane. Similarly, a measured bound state for Helix 1 Reporter (located at residue 4) would require the beginning of the first helix to engage with the membrane.

Therefore, the decrease in binding of Helix 1 over Helix 2 across the PD-associated mutants indicates that while Helix 1 may be critical in triggering binding of WT α-synuclein, the second helix plays a more dominant role over Helix 1 in associating the protein with the membrane for the PD-associated mutants.

The uniformity of the equilibrium shift observed across all PD-associated mutants suggests that a state in which the second α-helix is more tightly bound is related to the dysfunction of α-synuclein and may potentially serve as an initiator for fibrillization and eventual aggregation of α-synuclein. While this work does not extend to uncover how this state promotes aggregation, plausible mechanisms include 1) the solution-exposed first α-helix acting as a nucleation site, 2) the second α-helix acting as a nucleation site for membrane-templated aggregation, or 3) the shifted equilibrium alters binding partners of α-synuclein. To the best of our knowledge, there have been no comprehensive lipid-based aggregation studies on these PD-associated mutants, though several studies have identified the importance of solution-exposure of the N-terminus of α-synuclein in promoting fibrillization.[19,20]

Our findings reveal a uniform effect of PD-associated mutants on the binding of α-synuclein to lipid membranes. We highlight the utility of using an appropriate binding model in quantifying α-synuclein's binding to membranes with limited binding site availability – a lipid depletion regime that is often accessed within α-synuclein binding studies due to the protein's interactions with a large set of lipid parameters. Altogether, our work suggests that a bound state in which the second helix of α-synuclein is more tightly bound than the first may play a critical role within PD-progression and warrants future exploration for its role in protein aggregation and dysregulation.

4.9. α-Synuclein: Future Direction

Our lipid-binding analyses point to differences in membrane-engagement between the two helices of α-synuclein, suggesting that this may be a critical feature of either α-synuclein's normal function or to the progression of disease-associated states. While we have identified the uniform effect of the PD-associated mutations on shifting the equilibrium bound state of α-synuclein towards the second helix, our work has not yet extended to uncovering if this state plays a role within the aggregation of α-synuclein and disease progression. Using a ThT binding assay, we can correlate the rate of fibril formation with the relative dominance of the Helix 2 bound state for each PD-associated mutation within the lipid depletion regime.[21] Paired with dynamic light scattering (DLS),[22] we can further identify how a Helix 2-dominant bound state alters fibril size. To further verify the role of this state in fibril formation, we can use the amphipathic properties of the KTKEGV motifs found along the helices of α-synuclein to engineer α-synuclein variants with an intentionally stabilized second helix over the first.[23] Interestingly, several of the mutations that would alter the helix register shift in favor of the second helix, such as V40G, E61K, and Q79R, have been shown to increase the toxicity of α-synuclein.[24,25] By systematically examining the α-synuclein variants' equilibrium bound states in parallel with their fibril rate formation and size, we can connect lipid-binding properties of α-synuclein with its role in neuronal dysfunction.

4.10. References

1. Klein, C. & Westenberger, A. Genetics of Parkinson's Disease. *Cold Spring Harb Perspect Med* **2**, a008888–a008888 (2012).

2. Whittaker, H. T., Qui, Y., Bettencourt, C. & Houlden, H. Multiple system atrophy: genetic risks and alpha-synuclein mutations. *F1000Res* **6**, 2072 (2017).

3. Ramalingam, N. & Dettmer, U. Temperature is a key determinant of alpha- and beta-synuclein membrane interactions in neurons. *Journal of Biological Chemistry* **296**, 100271 (2021).

4. Lautenschläger, J. *et al.* C-terminal calcium binding of α-synuclein modulates synaptic vesicle interaction. *Nat Commun* **9**, 712 (2018).

5.   Middleton, E. R. & Rhoades, E. Effects of Curvature and Composition on α-Synuclein Binding to Lipid Vesicles. *Biophys J* **99**, 2279–2288 (2010).

6.   Ruf, V. C. *et al.* Different Effects of α-Synuclein Mutants on Lipid Binding and Aggregation Detected by Single Molecule Fluorescence Spectroscopy and ThT Fluorescence-Based Measurements. *ACS Chem Neurosci* **10**, 1649–1659 (2019).

7.   Guan, Y. *et al.* Pathogenic Mutations Differentially Regulate Cell-to-Cell Transmission of α-Synuclein. *Front Cell Neurosci* **14**, (2020).

8.   Siddiqui, I. J., Pervaiz, N. & Abbasi, A. A. The Parkinson Disease gene SNCA: Evolutionary and structural insights with pathological implication. *Sci Rep* **6**, 24475 (2016).

9.   Srinivasan, B. A guide to the Michaelis–Menten equation: steady state and beyond. *FEBS J* **289**, 6086–6098 (2022).

10.   Sarchione, A., Marchand, A., Taymans, J.-M. & Chartier-Harlin, M.-C. Alpha-Synuclein and Lipids: The Elephant in the Room? *Cells* **10**, 2452 (2021).

11.   Afitska, K., Fucikova, A., Shvadchak, V. V. & Yushchenko, D. A. α-Synuclein aggregation at low concentrations. *Biochimica et Biophysica Acta (BBA) - Proteins and Proteomics* **1867**, 701–709 (2019).

12.   Poudel, K. R. & Bai, J. Synaptic vesicle morphology: a case of protein sorting? *Curr Opin Cell Biol* **26**, 28–33 (2014).

13.   Pfefferkorn, C. M., Jiang, Z. & Lee, J. C. Biophysics of α-synuclein membrane interactions. *Biochimica et Biophysica Acta (BBA) - Biomembranes* **1818**, 162–171 (2012).

14.   Jarmoskaite, I., AlSadhan, I., Vaidyanathan, P. P. & Herschlag, D. How to measure and evaluate binding affinities. *Elife* **9**, (2020).

15.   Stöckl, M., Fischer, P., Wanker, E. & Herrmann, A. α-Synuclein Selectively Binds to Anionic Phospholipids Embedded in Liquid-Disordered Domains. *J Mol Biol* **375**, 1394–1404 (2008).

16.   Rovere, M. *et al.* E46K-like α-synuclein mutants increase lipid interactions and disrupt membrane selectivity. *Journal of Biological Chemistry* **294**, 9799–9812 (2019).

17.   Bodner, C. R., Maltsev, A. S., Dobson, C. M. & Bax, A. Differential Phospholipid Binding of α-Synuclein Variants Implicated in Parkinson's Disease Revealed by Solution NMR Spectroscopy. *Biochemistry* **49**, 862–871 (2010).

18.   Bartels, T. *et al.* The N-Terminus of the Intrinsically Disordered Protein α-Synuclein Triggers Membrane Binding and Helix Folding. *Biophys J* **99**, 2116–2124 (2010).

19.   McGlinchey, R. P., Ni, X., Shadish, J. A., Jiang, J. & Lee, J. C. The N terminus of α-synuclein dictates fibril formation. *Proceedings of the National Academy of Sciences* **118**, (2021).

20. Stephens, A. D. *et al.* Extent of N-terminus exposure of monomeric alpha-synuclein determines its aggregation propensity. *Nat Commun* **11**, 2820 (2020).

21. Kurochka, A. S., Yushchenko, D. A., Bouř, P. & Shvadchak, V. V. Influence of Lipid Membranes on α-Synuclein Aggregation. *ACS Chem Neurosci* **12**, 825–830 (2021).

22. Li, X. *et al.* Early stages of aggregation of engineered α-synuclein monomers and oligomers in solution. *Sci Rep* **9**, 1734 (2019).

23. Dettmer, U. Rationally Designed Variants of α-Synuclein Illuminate Its in vivo Structural Properties in Health and Disease. *Front Neurosci* **12**, (2018).

24. Volles, M. J. & Lansbury, P. T. Relationships between the Sequence of α-Synuclein and its Membrane Affinity, Fibrillization Propensity, and Yeast Toxicity. *J Mol Biol* **366**, 1510–1522 (2007).

25. Kim, T.-K. *et al.* Inflammation promotes synucleinopathy propagation. *Exp Mol Med* **54**, 2148–2161 (2022).

CHAPTER 5

CHARACTERIZING THE PROTEIN CONCENTRATION DEPENDENCE OF THE LIPID-BINDING BEHAVIOR OF α-SYNUCLEIN

5.1. Introduction

Multiple pieces of evidence point to that the lipid-binding behavior of α-synuclein is likely dependent on protein concentration, including literature precedent for aggregation propensity[1] and our results detailed in **Chapter 4**. It is no surprise that α-synuclein binding is often plotted as the bound fraction vs. lipid:protein ratio, but cross-study comparison of protein-binding behavior is difficult due to α-synuclein's sensitivity to buffer and lipid conditions. Within **Chapter 4**, tryptophan fluorescence was used to quantify α-synuclein:lipid interactions within the nM protein concentration range (50-300 nM for tryptophan fluorescence) and our results already indicate that the protein's binding regime lies within depletion. As an extension of this work, we were curious how protein concentration dependence alters the effective binding site availability on the lipid membrane along with distribution of equilibrium bound states of WT α-synuclein.

To extend the experimentally-available protein concentration range, we explored a range of techniques to quantify α-synuclein:lipid binding. This included Fluorescence Polarization (FP) measurements for the ~40 nM-100 nM protein concentration range, and CD and ^{1}H-^{15}N HSQC NMR for the μM protein concentration range. The following sections discuss preliminary α-synuclein binding data for these methods along with their benefits and limitations within future work. Additionally, the sections briefly compare the results from the various techniques and contextualize them within the depletion model and the broader α-synuclein field.

5.2. α-Synuclein: Fluorescence Polarization

The tryptophan fluorescence assay can be used to explore a variety of parameters in α-synuclein:membrane interactions, but the technique is both material-consuming and time-consuming, where a single binding curve would require ~5-6 hours. A fluorescence polarization (FP) assay can be used to address these limitations while achieving a similar protein concentration range. The assay requires much less protein and lipid, as the sample size is ~150 μL, and ~6 binding curves can be assembled and scanned within an hour. Similar to tryptophan fluorescence, FP can explore a variety of protein, lipid, and buffer conditions and is suitable for comparing α-synuclein mutants.[2,3] However, a measured "bound" state by FP is not site-specific since an α-synuclein bound to a lipid vesicle will have a decreased tumbling rate regardless of bound state, and it therefore cannot be used to identify differences in the bound state of the two α-helices in α-synuclein. Regardless, the method still has utility in measuring overall α-synuclein binding.

Within the work here, our aim was to validate FP as a method to quantify α-synuclein:lipid binding through a comparison to tryptophan fluorescence. As a part of this investigation, we explored how fluorophore placement affected measured binding, compared measured overall binding of α-synuclein to helix-specific binding data from tryptophan fluorescence, and verified FP as a method to report on *relative* affinity and binding site density of the PD-associated A30P and E46K mutation. A portion of this work was completed in collaboration with Haley Sturgill, a summer high school student, who contributed both intellectually and carried out several of the experiments used within this study.

5.2.1. α-Synuclein FP: Fluorophore Placement

FP measurements were carried out with α-synuclein tagged using the Alexa488 fluorophore. Due to its size and hydrophobicity, a concern was that the fluorophore could potentially impede α-helix formation or promote membrane:protein interactions between the aromatic groups of the fluorophore and lipid tails. To address this concern, we expressed α-synuclein with cysteine mutations at multiple sites, S9C, S42C, G93C, that were each subsequently tagged with the fluorophore. We then carried out the FP measurements and compared them to binding data from our tryptophan fluorescence assays.

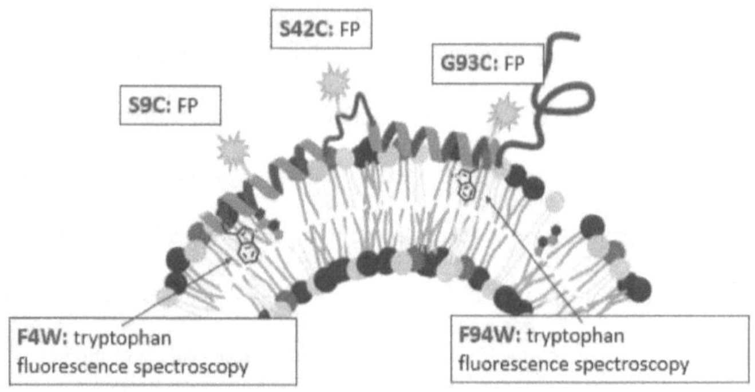

Fig 5.1. Comparison of tryptophan mutations (F4W and F94W) and sites for fluorophore placement (S9C, S42C, and S93C) for tryptophan fluorescence spectroscopy and fluorescence polarization.

We found that tagging at site G93C impeded binding the least out of the three sites tested (**Fig. 5.1**), where qualitatively, the binding curve for G93C best aligns with the binding curve from tryptophan fluorescence of α-synuclein. Fit parameters as well aligned between the two methods (**Table 5.1**), albeit with much larger confidence intervals from FP. Tagging at sites S9C and S42C (data not shown) appeared to have, respectively, a minor and large decrease on the affinity of α-synuclein for the lipid membrane, likely due either directly impediment of helix formation or prevention of membrane association.

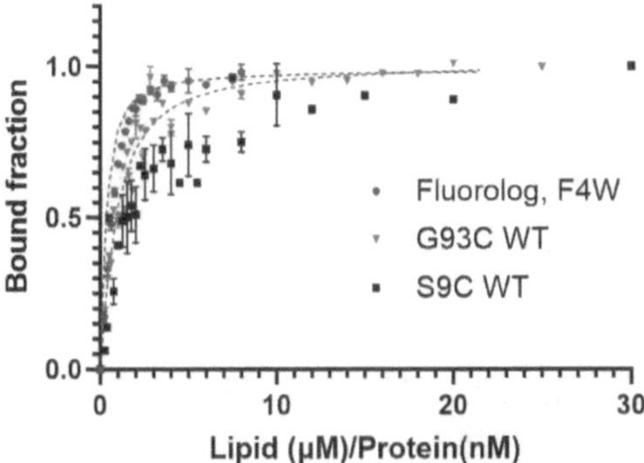

Fig. 5.2. Binding curves of WT α-synuclein to lipid vesicles composed of 55:20:15:10 DOPC:DOPS:DOPE:Chol with a diameter of ~70 nm tagged with Alexa488 at the S9C (black) or G93C sites (green). Data were fit to **Eq. 4.1** and best fit, where applicable, is shown as a dashed line. Reference binding curve from tryptophan fluorescence spectroscopy (F4W site) is shown in red.

Of note, tryptophan fluorescence and FP experiments were completed under two separate protein concentrations (125 nM and 40 nM, respectively) and it is possible that the lower protein concentrations of the FP experiments represent a binding regime that is further outside the depletion regime. This could potentially explain why the FP curve aligns with the tryptophan fluorescence curve at lower lipid concentrations but begins to deviate closer to saturation where the protein's binding regime is better represented by the effect of the K_d. While we tested varying protein concentrations for FP, we found that signal noise became too high above 100 nM protein concentrations, which could be a result of too high of a total fluorophore concentration. Future work can involve varying the ratio of labeled to unlabeled α-synuclein under constant total protein concentration to extend the experimentally-available binding regimes to >100 nM. The higher noise at higher protein concentrations for FP may also be explained by increased protein:protein interactions. For example, the slower tumbling speed of an α-synuclein dimer or multimer would increase the anisotropy, resulting in non-binary contributions of α-synuclein states to the total measured anisotropy.

Table 5.1. Fit parameters for binding curves generated by tryptophan fluorescence (F4W) and FP (S9C and G93C) of WT α-synuclein to lipid vesicles composed of 55:20:15:10 DOPC:DOPS:DOPE:Chol with a diameter of ~70 nm. Binding curves were fit using **Eq. 4.1** with a protein concentration of 125 nM for tryptophan fluorescence and 40 nM for FP. Data are displayed as 95% confidence intervals except for values denotes by [1], for which confidence intervals could not be calculated. Parameters that could not be fit are denoted by an *.

	F4W	FP S9C	FP G93C
σ (1/lipids)	[0.0011, 0.0012]	*	[0.0008729, 0.00187]
lipids (1/σ)	[833, 909]	*	[533, 1146]
K_d (μM)	[20.3, 27.5]	60.5[1]	[14.0, 39.5]
B_{max}	[0.9987, 1.007]	[0.9150, 1.0664]	[0.9661, 1.073]
R^2	0.9987	0.92	0.958

The protein concentration range of tryptophan fluorescence assays is largely limited by the increased lipid scattering above ~3 mM total lipid, and for higher protein concentrations it is likely that saturation of binding cannot be observed. In using FP, we were concerned that the high lipid vesicle concentrations may lower the measured anisotropy due to vesicle scattering. As bound fractions are determined from the anisotropy of a fully bound state, this could artificially increase the calculated bound fraction for intermediate lipid concentrations. While ideally we would determine the effects of scattering through a negative control, we do not have a suitable lipid composition to which α-synuclein does not bind, as we have observed modest binding even to vesicles composed of 100% POPC (data not shown). However, within our measured curves, anisotropy values do not decrease following plateau, suggesting that the higher lipid concentrations do not have a large effect on the measurements within the lipid concentrations used within this study.

5.2.2. α-Synuclein FP: PD-associated Mutants, A30P and E46K

To further explore and validate FP as a method to quantify the binding of α-synuclein to lipid membranes, we measured the binding of α-synuclein with the PD-associated mutations, A30P and E46K, tagged at the G93C site. As discussed in **Chapter 4**, within the depletion regime, there is higher binding site availability for the second helix of the PD-associated mutants over the first helix. We expected that for FP, a method that measures global protein binding (see discussion in methods, **Chapter 3.2.2**), the binding curve and its parameter fits would more closely align with the tryptophan fluorescence binding curves for the F94W (Helix 2) probe.

As shown in **Fig. 5.3**, we see that the FP binding curves for the E46K and A30P mutations follow the expected trend with the A30P mutant having a lower affinity for lipid than E46K (see Fig. 4.7). However, the binding curves for both mutants deviate down from the tryptophan fluorescence curves of their respective binding curve from the Helix 2 probe at higher lipid concentrations. This deviation is greater than the observed deviation for WT α-synuclein (**Fig. 5.2**).

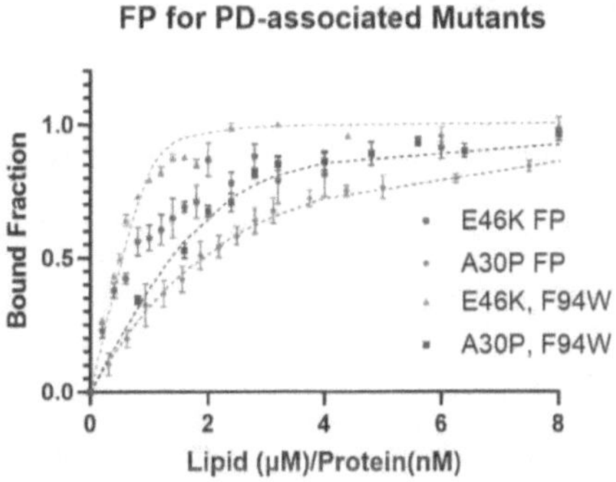

Fig. 5.3. Binding curves of α-synuclein with the PD-associated mutations A30P and E46K to lipid vesicles composed of 55:20:15:10 DOPC:DOPS:DOPE:Chol with a diameter of ~70 nm using tryptophan fluorescence spectroscopy or FP. Tryptophan fluorescence curves are shown for the F94W (Helix 2) probe and were taken at a protein concentration of 125 nM. FP curves are shown for α-synuclein tagged at G93C and were taken at a protein concentration of 40 nM. The bound fraction of all curves is plotted against lipid:protein ratio to allow for comparison between the two protein concentrations used between the methods, and curves are truncated to better show detail at

lower lipid concentrations (all binding curves reach plateau). Bound fractions and errors bars correspond to the averages and standard error of 3-5 trials at each lipid concentration. Curves of best fit (**Table 5.2**) for each binding curve are shown as dashed lines except for the FP binding curve for the E46K mutant.

Both techniques observed a false plateau at around ~60-70% of protein bound (**Fig. 5.2**), a feature that has similarly appeared in binding data across other studies[4,5]. We were unsure if this false plateau is due to direct changes in the protein's bound state with increased lipid availability, such as decreased protein:protein interactions or differences in protein conformational state (e.g., bent to straight[6]), or if the plateau was due to changes in the distribution of binding site occupancies as the protein shifted from less preferred (lower affinity) to more preferred (higher affinity) binding sites.

It is possible that the signal corresponding to a "bound" protein may be different across states. For example, the arrangement of lipids that comprises a binding site may alter tryptophan fluorescence emission intensity due to increased or decreased hydration. In such cases, normalizing an intermediate lipid concentration to a fully saturated state may result in an under or over estimation of bound fraction of protein, yielding a false plateau in the region where protein distribution shifts from one state to another. However, it is unlikely that two separate techniques would both display a similar feature in the binding curve, suggesting that there is a true change in total bound protein at around ~70% protein bound.

The false plateau may indicate a change in the distribution of bound states across α-synuclein's binding curve with increased lipid availability. This possibility in conjunction with the observed changes between WT α-synuclein and the PD-associated mutants suggests that the G93C site affects the affinity of some bound states but not all, explaining the only modest deviation from the tryptophan fluorescence curve for WT but much greater deviations for the PD-associated mutants.

At lower lipid concentrations, the binding curves for WT α-synuclein and the E46K mutant qualitatively align between the two methods, while the A30P mutant deviates much earlier. As both WT and the E46K mutant have higher affinity for lipid membranes, the earlier deviation of the A30P mutant may be explained by the protein exiting the depletion regime earlier and becoming sensitive to the effects of tagging at the G93C site. This possibility is supported by the parameter fits of the A30P binding curve, where σ_{FP} is closer to σ_{H2} but the K_d fit for FP is greater than tryptophan fluorescence for the Helix 2 probe (**Table 5.2a**).

We were unable to fit the binding curve for the E46K mutant (**Table 5.2b**), likely due to the false plateau at ~60% bound fraction. While this dip was reproducible across several trials, values for the bound fraction fluctuated more within this region than for lower or higher lipid concentrations. Due to α-synuclein's propensity to aggregate during purification, it is unclear if this dip is a result of the protein preparation or the potential for the combined effects of the E46K mutation and the fluorophore at G93C to alter α-synuclein's bound or solution conformational states. Additional protein purifications can help clarify the extent of this dip's dependence on protein preparation and stock concentrations.

Table 5.2a. Fit parameters for binding curves generated by tryptophan fluorescence (F4W and F94W) and FP (G93C) of the PD-associated mutants **a.** A30P and **b.** E46K α-synuclein to lipid vesicles composed of 55:20:15:10 DOPC:DOPS:DOPE:Chol with a diameter of ~70 nm. Binding curves were fit using **Eq. 4.1** with a protein concentration of 125 nM for tryptophan fluorescence and 40 nM for FP. Data are displayed as 95% confidence intervals except for values denotes by [1], for which confidence intervals could not be calculated. Parameters that could not be fit are denoted by an *.

	FP A30P	A30P F4W	A30P F94W
σ (1/lipids)	[0.0005353, 0.00108]	[0.00032, 0.00045]	[0.00036, 0.00061]
lipids (1/σ)	[927, 1868]	[2341, 3086]	[1671, 2635]
K_d (µM)	[39.5, 61.4]	[5.2, 50]	[12.8, 63]
B_{max}	[1.000, 1.043]	[0.9483, 1.036]	[0.9256, 1.023]
R^2	0.99	0.9924	0.9899

Table 5.2b, continued.

	FP E46K	E46K F4W	E46K F94W
σ (1/lipids)	*	[0.0072, 0.00091]	[0.0012, 0.0027]
lipids (1/σ)	*	[1070, 1362]	[470, 681]
K_d (μM)	31.7*	[1.1, 14.5]	[11.44, 27.1]
B_{max}	[0.9784, 1.128]	[0.9735, 1.065]	[0.9538, 1.073]
R^2	0.97	0.9953	0.9887

5.2.3. α-Synuclein FP: Conclusions

Overall, we conclude that fluorophore tagging of the G93C site can adequately quantify α-synuclein's interactions with lipid at lower lipid concentrations and nanomolar protein concentrations, and FP can be used as a method to compare relative differences between protein or lipid conditions within the depletion regime. However, it appears that FP should be used with caution as preliminary data suggest that the fluorophore may have a modest effect on the affinity of the protein for membranes for some states, advising for data to be analyzed in conjunction with binding data from methods with helix-specific resolution.

5.3. α-Synuclein: CD

Circular dichroism allows us to probe at changes in the structural features of α-synuclein across the lipid titration.[3] The method has previously been used to identify transitions of α-synuclein from its disordered state to its α-helical structure in the presence of lipid membranes and can be used as an approximation for quantifying membrane-binding.[7–9]

We performed CD measurements of F4W α-synuclein at a protein concentration of 6 μM for a lipid concentration range of 0 μM to 15,000 μM. In our tryptophan fluorescence experiments, binding saturates at ~ 250 μM lipid for a protein concentration of 125 nM. Assuming a constant protein:lipid saturation ratio between the two methods, we would expect binding to saturate at ~12,000 μM lipid for 6 μM protein. Similarly, 70% of protein bound would correspond to an approximate lipid concentration of ~6,000 μM. Of note, we do not necessarily expect the protein:lipid saturation ratio to be constant between the two protein concentrations as they represent different binding regimes for α-synuclein, but the direct comparison does establish a rough expectation for protein saturation of the lipid membrane.

CD spectra of α-synuclein show a gradual transition from disordered structure to increased α-helical content with the addition of lipid with a maximal $[\Theta]_{222}$ value at 4 mM lipid. Following this, we see a decreased absolute value for ellipticity, likely due to vesicle scattering of the high lipid concentrations.[9] These spectra align with α-synuclein's proposed mechanism of binding, with the protein transitioning from its intrinsically disordered state in solution to the of formation of two α-helices upon binding to membrane (see **Fig. 2.5**).

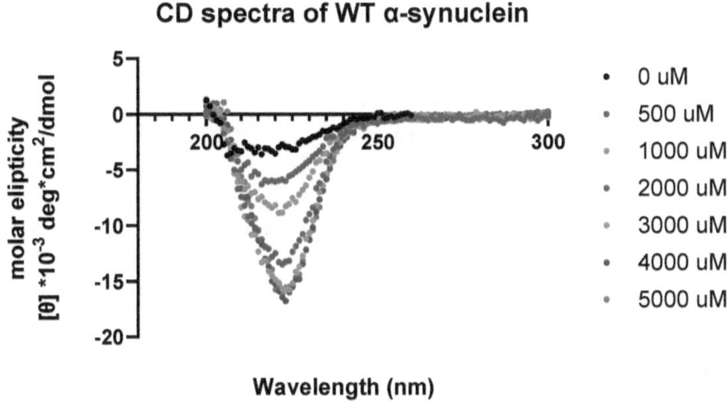

Fig. 5.4. CD spectra of lipid titration of 6 μM α-synuclein to lipid vesicles composed of 55:20:15:10 DOPC:DOPS:DOPE:Chol with a diameter of ~70 nm. The absolute values of spectra corresponding to higher lipid concentrations begin to decrease, likely due to scattering,[9] and are therefore not shown.

While we cannot deduce the site-specificity of α-helicity from these data, we can use the maximal helicity to approximate the fraction bound of α-synuclein at intermediate lipid concentrations. However, it is important to note that the fraction bound is not necessarily identical to the fraction of protein in its α-helical state and one should be careful in interpreting the data and should use them as only approximations. For example, at an intermediate lipid concentration, the bound state of α-synuclein may favor the binding and formation of the second helix, resulting in an underestimate of the total protein bound to the membrane. The approximation of the binding curve is shown in **Fig. 5.5**.

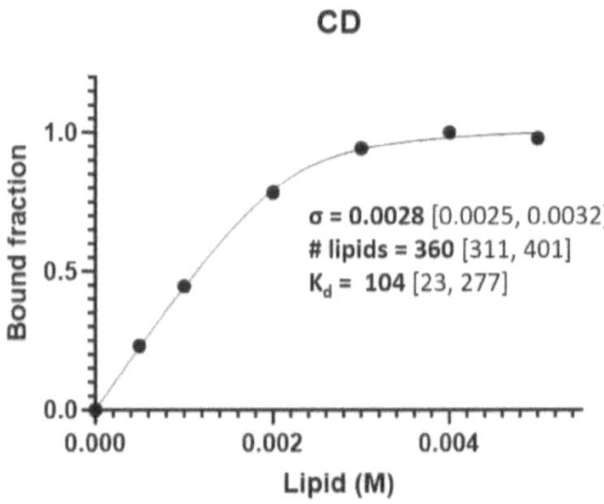

Fig. 5.5. Approximate binding curve of 6 μM α-synuclein to lipid vesicles composed of 55:20:15:10 DOPC:DOPS:DOPE:Chol with a diameter of ~70 nm. Binding data were fit using **Eq. 4.1** using a $[P]_{tot}$ = 6 μM with fit parameters shown in figure legend.

The binding curve from the CD data shows a clear linear trend at lower lipid concentrations, corresponding to a binding site density of 1 protein per 360 lipids (compared to a binding site density of 1 per ~909 lipids at the 125 nM concentrations of tryptophan fluorescence). A more thorough analysis of protein concentration dependence on the binding of α-synuclein to lipid membranes can be found in **Section 5.5**, but the CD spectra suggest that the binding regime differs between the concentrations used in the two techniques and lies further within the depletion regime. Altogether, CD can be used as an approximation for quantifying lipid binding but likely cannot give state-specific resolution of the protein's bound states on the lipid membrane.

5.4. α-Synuclein: ^{1}H-^{15}N HSQC NMR

5.4.1. α-Synuclein: ^{1}H-^{15}N HSQC NMR, WT

Quantification of α-synuclein:lipid measurements at higher μM protein concentrations can be achieved through ^{1}H-^{15}N HSQC NMR experiments.[10] This technique has the additional benefit of having helix-specific resolution along with the ability to track protein conformational changes across conditions. With this technique, we aimed to compare the availability of lipid binding sites for the two helices of α-synuclein between the nM protein concentrations of tryptophan fluorescence spectroscopy and the μM protein concentrations of ^{1}H-^{15}N HSQC NMR. These comparisons may help clarify the discrepancies between the reported binding values in literature across various protein concentration.

We first expressed ^{15}N-labeled α-synuclein with the Dual Reporter (F4W, F94W) for a direct comparison to binding data from tryptophan fluorescence. The F4W and F94W mutations additionally have the benefit of simplifying peak identification and integration, as peaks corresponding to the Hε-Nε of the sidechains of tryptophan residues are found in a different region of the ^{1}H-^{15}N HSQC NMR spectrum than the peaks corresponding to the amide backbone of a protein.[11]

The ^{1}H-^{15}N HSQC NMR spectra of 50 μM protein in the absence of lipid show even peaks corresponding to F4W and F94W (**Fig. 5.6a**). However, the addition of 10.44 mM lipid results in the complete disappearance of the F94W peak and the partial disappearance of the F4W peak (**Fig. 5.6b**). At the intermediate 5.22 mM lipid concentration, the F94W peak is similarly not present, but the F4W peak is stronger than that at 10.44 mM lipid, indicating less binding of the F4W residue to the membrane at lower lipid concentration. In comparison, the peak corresponding to residue A140 remains constant in all three conditions (0, 5.22, and 10.44 mM lipid). Residue A140 is located on the C-terminus of α-synuclein, which, as discussed in **Chapter 2.2.1**, does not directly

interact with the lipid membrane and projects off the surface, and can therefore be used as a control

for protein concentration and scan conditions. Volume integration for these peaks under the three

conditions are shown in **Table 5.3** along with the values adjusted to the volume of the A140 peak.

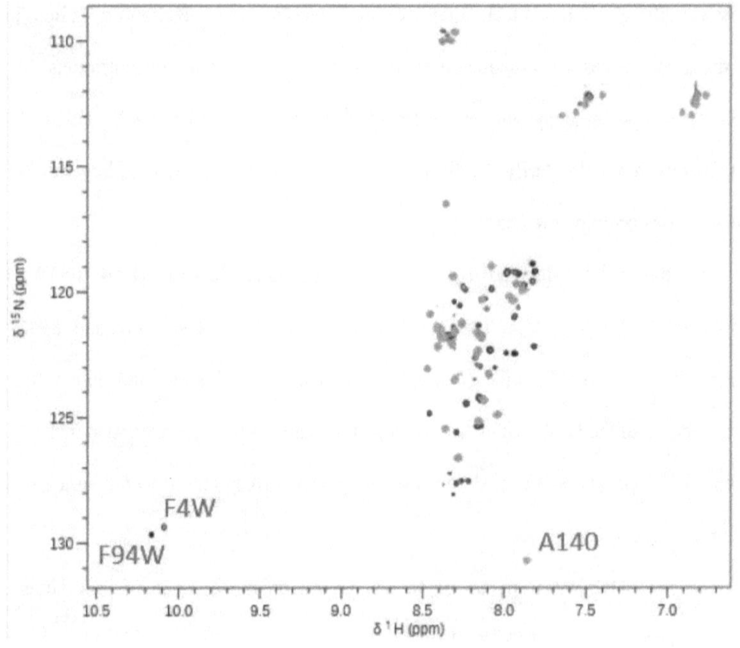

Fig. 5.6. ¹H-¹⁵N HSQC NMR spectra of WT α-synuclein.

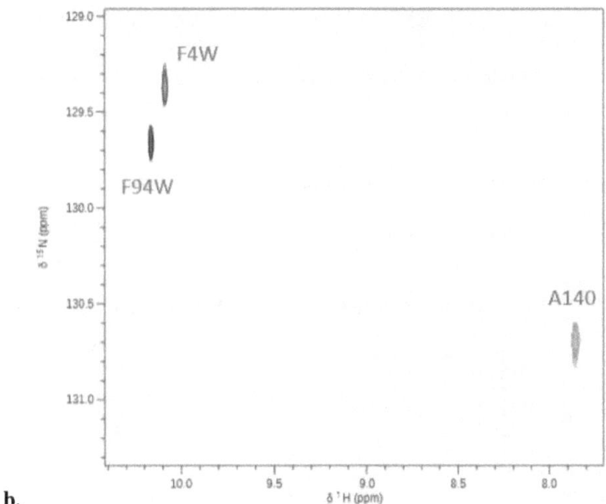

b.

Fig. 5.6, continued. a. ^{1}H-^{15}N HSQC NMR spectra of 50 µM α-synuclein with Dual Reporter (F4W, F94W) in the absence of lipid (black) overlaid by samples in the presence of 5.22 mM (blue) and 10.44 mM (red) lipid vesicles composed of 55:20:15:10 DOPC:DOPS:DOPE:Chol with a diameter of ~70 nm. Peaks corresponding to Hε-Nε of the sidechains of F4W and F94W are in the bottom left corner. Peak corresponding to A140, as identified by BMRB entry 18857, can be found in the bottom right. **b.** Zoom-in of the three identified peaks, showing the fast and gradual disappearance of the F94W and F4W peaks, respectively, and the continued presence of the A140 peak.

Table 5.3. Peak volumes of A140, F4W, and F94W residues of WT α-synuclein spectra in **Fig. 5.6**, as found through the peak pick and integration tool of **NMRFx**. Adjusted values for the F4W and F94W peaks.

	0mM lipid	**5.22 mM lipid**	**10.44 mM lipid**
A140	0.99201 +/- 0.00942	1.01262 +/- 0.00982	0.97763 +/- 0.01057
F4W	0.68764 +/- 0.00872	0.17024 +/- 0.00743	0.07933 +/- 0.00647
F94W	0.65583 +/- 0.00872	0	0
F4W, Adjusted	0.69318 +/- 0.0128	0.16812 +/- 0.0123	0.08115 +/- 0.0124
F94W, Adjusted	0.66111 +/- 0.0128	0	0

The disappearance of the F4W and F94W peaks can be used to track and quantify the binding of

α-synuclein to lipid membranes with site-specificity. For this, we measured a lipid-binding curve

for α-synuclein, and for each lipid condition, we used the A140-adjusted F4W and F94W peak

volumes to determine the unbound fraction at each lipid concentration. The binding curves for

both sites are shown in **Fig. 5.4**.

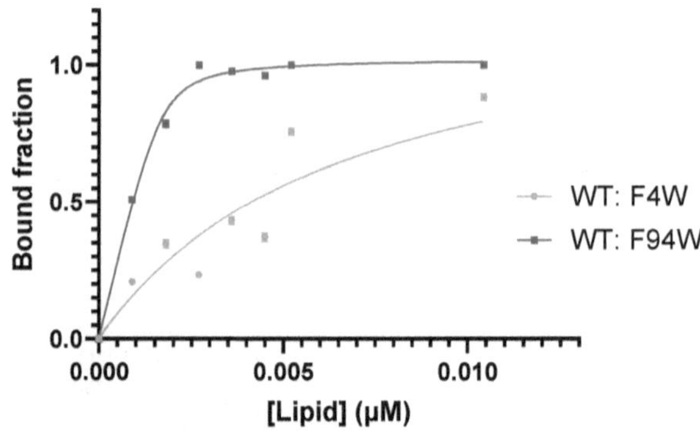

Fig. 5.7. Binding curves for WT α-synuclein for the F4W and F94W sites as found through ^{1}H-^{15}N HSQC NMR fit to **Eq. 4.1**. Error bars represent estimated error for peak volume of one replicate of the experiment, as determined using NMRFx software.

Both the raw spectra and the binding curves for WT α-synuclein indicate that the F94W site is

more tightly associated with the membrane than the F4W site – a stark difference from the

observed bound states of the protein from tryptophan fluorescence spectroscopy. Even more

interestingly, we see that the apparent binding site availability for both sites as determined through

NMR, is much greater than the binding site availability found from tryptophan fluorescence (**Table**

5.4) with a 25x decrease in the approximate number of lipids per binding site for site F94W and a

6x decrease for the F4W site.

Table 5.4. Fit parameters for binding curves generated by NMR for the F4W and F94W sites for WT α-synuclein to lipid vesicles composed of 55:20:15:10 DOPC:DOPS:DOPE:Chol with a diameter of ~70 nm compared to fit parameters yielded from tryptophan fluorescence (TF). Binding curves were fit using **Eq. 4.1** with a protein concentration of 50 μM for NMR and 125 nM for tryptophan fluorescence. Data are displayed with 95% confidence intervals except for values denotes by [1], for which confidence intervals could not be calculated. Parameters that could not be fit are denoted by an *.

	NMR, F4W	NMR, F94W	TF (F4W)
σ (1/lipids)	0.006579[1]	0.029 [0.0218, 0.0531]	0.0011 [0.00011, 0.0012]
lipids (1/σ)	152[1]	34 [18.8, 45.9]	909 [833, 909]
K_d (μM)	*	97 [15.5, 423]	23.9 [20.3, 27.5]
B_{max}	1.295[1]	1.022 [0.9602, 1.116]	0.999 [0.9987, 1.007]
R^2	0.84	0.993	0.9987

NMR and tryptophan fluorescence spectroscopy similarly represent site-specific association of residues with the membrane. While NMR does not necessitate direct insertion of the residues, it does require the rotational freedom of the residue to be limited by the tumbling speed of the vesicle for a residue to be considered "bound." Therefore, the differences between the F4W and F94W site are likely real differences in the bound (**Fig. 5.5**) states of the protein within the NMR samples that were not observed in the tryptophan fluorescence spectroscopy samples.

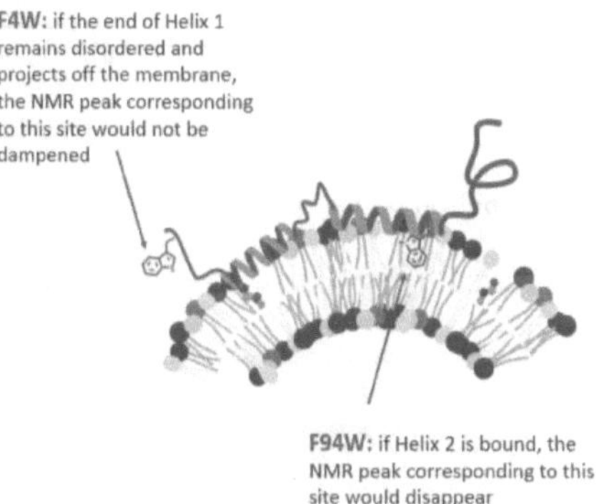

Fig. 5.8. Graphic representation of a bound protein state in which the end of Helix 1 remains unbound, corresponding to the retention of its NMR peak.

It is possible that NMR observes membrane-associated states that tryptophan fluorescence spectroscopy does not. For example, within membrane-templated fibrillization an NMR peak of a residue in a membrane:oligomer complex would disappear, but the corresponding tryptophan could remain solution-exposed and would not experience a shift in its emission spectrum. However, FP binding data would reflect these states, and with close alignment in fit parameters between FP and tryptophan fluorescence spectroscopy, it is unlikely that the binding data found from NMR can be entirely attributed to differences in the methods' definitions of "bound" states. Overall, the methods have similar buffer, temperature, and lipid conditions but very different protein concentration ranges (50 μM vs. 40-125 nM), suggesting that the binding of α-synuclein to lipid membranes is highly dependent on protein concentration.

5.4.2. α-Synuclein: ^{1}H-^{15}N HSQC NMR: G51D Mutant

We performed a similar preliminary analysis of ^{1}H-^{15}N HSQC NMR spectra corresponding to lipid-binding of α-synuclein with the PD-associated mutation G51D (**Fig. 5.6**), where we found a similar gradual disappearance of the F4W peak with a faster disappearance of the F94W peak. The comparison of WT α-synuclein to the G51D mutant binding curves of the two peaks are shown in **Fig. 5.7** and the corresponding fits are shown in **Table 5.5**. Qualitatively, while the binding of the F4W and the F94W sites appear to be similar between WT α-synuclein and the G51D mutant (contrasting with the results of **Chapter 4**), we cannot reasonably analyze the lipid-binding region of the F94W site as it is under sampled. However, the work here demonstrates that this technique can be reasonably used to further resolve differences in the helix-dependent binding of the PD-associated mutants at μM protein concentrations.

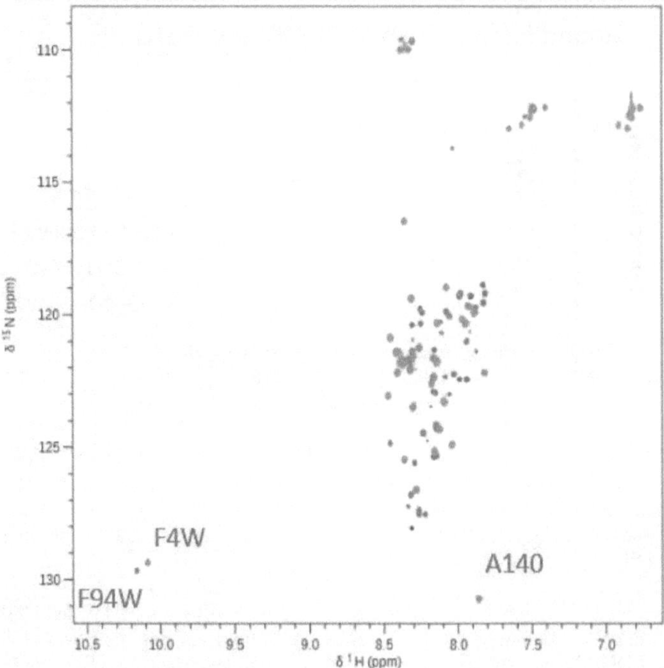

Fig. 5.9. a. ^{1}H-^{15}N HSQC NMR spectra of 50 µM α-synuclein with the PD-associated mutation G51D and the Dual Reporter (F4W, F94W) with no lipid (black) overlaid by samples in the presence of 0.9 mM (blue) and 11.88 mM (red) lipid vesicles composed of 55:20:15:10 DOPC:DOPS:DOPE:Chol with a diameter of ~70 nm. Peaks corresponding to Hε-Nε of the sidechains of F4W and F94W are in the bottom left corner and peak corresponding to A140, as identified by BMRB entry 18857, can be found in the bottom right.

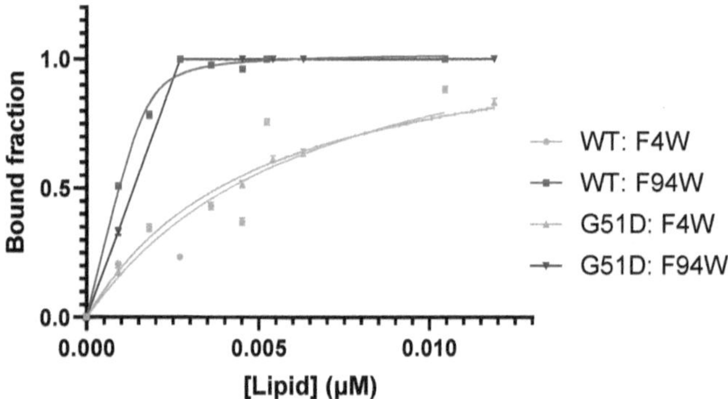

Fig. 5.10. Binding curves for the F4W and F94W sites from ^{1}H-^{15}N HSQC NMR of WT α-synuclein and the G51D mutant fit to **Eq. 4.1**. Error bars represent estimated error for peak volume of one replicate of the experiment, as determined using NMRFx software. Of note, the K_d region (near saturation) of the G51D, F94W curve is under sampled, as reflected by the sharp corner in its estimated fit.

Table 5.5. Fit parameters for binding curves generated by NMR for the F4W and F94W sites for WT and the PD-associated mutant G41D α-synuclein to lipid vesicles composed of 55:20:15:10 DOPC:DOPS:DOPE:Chol with a diameter of ~70 nm. Binding curves were fit using **Eq. 4.1** with a protein concentration of 50 µM. Data are displayed with 95% confidence intervals except for values denotes by [1], for which confidence intervals could not be calculated, or for the G51D, F94W curve, for which fitting yielded a perfect fit. Parameters that could not be fit are denoted by an *.

	WT, F4W	WT, F94W	G51D, F4W	G51D, F94W
σ (1/lipids)	0.006579[1]	0.029 [0.0218, 0.0531]	0.0049[1]	0.0186
lipids (1/σ)	152[1]	34 [18.8, 45.9]	204[1]	53.8
K_d (µM)	*	97 [15.5, 423]	*	~ 0
B_{max}	1.295[1]	1.022 [0.9602, 1.116]	1.146 [0.8357, 1.927]	1
R^2	0.84	0.993	0.997	1.00

5.5. Discussion of the Protein Concentration Dependence in α-Synuclein:Lipid Binding

Lipid binding data of α-synuclein differs between the µM protein concentrations of CD and ^{1}H-^{15}N HSQC NMR and the nM protein concentrations of tryptophan fluorescence spectroscopy and FP. Across these methods, we see large decreases in the lipid:protein ratio at which the protein reaches saturation with increased protein concentration. At the nM protein concentrations of tryptophan fluorescence spectroscopy and FP, α-synuclein reaches saturation at a lipid:protein ratio of ~2,000, while at the 6 μM and 50 μM concentrations of CD and NMR, α-synuclein binding is saturated, respectively, at ratios of ~750 and ~55-200 (lower and upper limits based on saturation of the F94W and F4W sites, respectively). The binding curves of WT α-synuclein as obtained using techniques with wildly varied protein concentrations – tryptophan fluorescence in the nM regime, CD and NMR in the μM regime – are displayed in **Fig. 5.8,** with fits of the data shown in **Table 5.6**. While we are not aware of any studies that directly bridge this protein concentration range under the same conditions, these values do align with published binding data at nearby concentration ranges on similar lipid compositions. For example, Makasewicz et al.[12] see saturation at a lipid:protein ratio of ~1000 on 7:3 DOPC:DOPS vesicles for 250 nM protein, while Fusco et al.[10] observe saturation at a ratio of ~65 for vesicles composed of 5:3:2 DOPE:DOPS:DOPC at 300 μM protein.

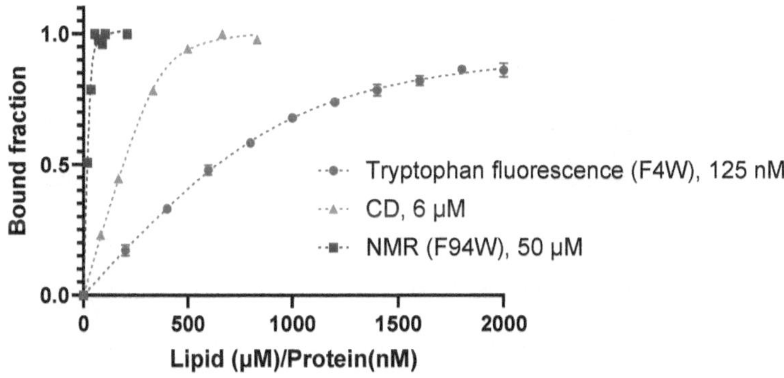

Fig. 5.11. Binding curves of α-synuclein to lipid vesicles composed of 55:20:15:10 DOPC:DOPS:DOPE:Chol with a diameter of ~70 nm at varying protein concentrations, using

tryptophan fluorescence spectroscopy (red), CD (green), and ^{1}H-^{15}N HSQC NMR (blue). Methods' respective protein concentrations are shown in the figure legend, and curves of best fit to **Eq. 4.1** for each data set are shown as dashed lines. Data are plotted against the protein:lipid ratio for ease of comparison between methods. Data for CD is shown for one replicate and does not contain error bars.

Table 5.6. Fit parameters for binding curves of WT α-synuclein to lipid vesicles composed of 55:20:15:10 DOPC:DOPS:DOPE:Chol with a diameter of ~70 nm compared to fit parameters yielded from tryptophan fluorescence (TF), CD, and NMR. Binding curves were fit using **Eq. 4.1** with a protein concentration of 125 nM for tryptophan fluorescence, 6 µM for CD, and 50 µM for NMR. Data are displayed with 95% confidence intervals except for values denotes by ^{1}, for which confidence intervals could not be calculated. Parameters that could not be fit are denoted by an *.

	TF (F4W)	CD	NMR, F94W
σ (1/lipids)	**0.0011** [0.00011, 0.0012]	**0.0028** [0.0025, 0.0032]	**0.029** [0.0218, 0.0531]
lipids (1/σ)	909 [833, 909]	360 [311, 401]	34 [18.8, 45.9]
K_d (µM)	23.9 [20.3, 27.5]	104 [23, 277]	97 [15.5, 423]
B_{max}	0.999 [0.9987, 1.007]	1.033 [0.981, 1.108]	1.022 [0.9602, 1.116]
R^2	0.9987	0.99	0.993

The protein-concentration dependence of the binding of α-synuclein to lipid membranes may be related to several key behaviors of the protein. The binding of α-synuclein to lipid membranes has been proposed to be cooperative,[12] where the saturation of the membrane of a single vesicle is preferred over the even distribution across all vesicles in the sample. It is possible that this cooperativity is only observed past a critical protein concentration if solution-based protein:protein interactions or solution-availability of α-synuclein are necessary for its binding mechanism.

α-synuclein has also been reported to disrupt the structure of lipid membranes. Previous reports have linked the formation of solution-based α-synuclein fibril structures to the disintegration of SV-mimic and other lipid membranes,[13] potentially providing a plausible link between the lipid-binding properties of α-synuclein and the formation of lipid-containing Lewy bodies. While our work does not currently extend to include the analysis of α-synuclein multimers, the aggregation propensity and dysfunction of α-synuclein has been clearly linked to the formation of fibrils and other higher order oligomer structures.[14] The presence of these structures and their impact on lipid-binding are also likely dependent on protein-concentration, warranting a deeper investigation into the concentration dependence of the solution state of α-synuclein and its lipid-binding properties.

5.6. Protein Concentration Dependence: Conclusions and Future Directions

Differences in the binding site availability of the two helices between the nM and μM protein concentration ranges may aid in understanding protein function and disease progression as both the PD-associated mutants (**Chapter 4**) and higher WT concentrations (**Chapter 5.4**) point to non-equal binding of the two α-helices. Through the three techniques shown in **Fig. 5.8**, we can cover a wide range of protein concentrations to identify critical concentrations at which the binding regime is altered to connect the features of α-synuclein's solution state to its lipid-bound state..Of note, all the work and cross-experiment comparisons within this book examined the binding of α-synuclein to lipid vesicles composed of 55:20:15:10 DOPC:DOPS:DOPE:Chol with a diameter of ~70 nm. While it is important to ensure that lipid samples are similar across experiments, it is also important to consider the dependence on binding site availability across various lipid and buffer conditions, as both factors likely influence the binding regime of the protein:lipid system.
Similarly, to further connect the conclusions drawn from our work with PD-associated mutants in **Chapter 4** and protein concentration dependence, it will be imperative to examine if similar differences are observed between WT α-synuclein and the PD-associated mutants at higher protein concentrations.
With the number of factors that appear to affect α-synuclein:membrane interactions, a thorough understanding α-synuclein's role within the neuron and disease progression will require comprehensive multi-dimensional analyses. Our work across **Chapter 4** and **Chapter 5**

demonstrate that these analyses can reveal details of the complex interplay between the bound

states of α-synuclein and relative binding site availability and protein concentration.

5.7. References

1. Afitska, K., Fucikova, A., Shvadchak, V. V. & Yushchenko, D. A. α-Synuclein aggregation at low concentrations. *Biochimica et Biophysica Acta (BBA) - Proteins and Proteomics* **1867**, 701–709 (2019).

2. Hall, M. D. *et al.* Fluorescence polarization assays in high-throughput screening and drug discovery: a review. *Methods Appl Fluoresc* **4**, 022001 (2016).

3. Galvagnion, C. *et al.* Lipid vesicles trigger α-synuclein aggregation by stimulating primary nucleation. *Nat Chem Biol* **11**, 229–234 (2015).

4. McClain, S. M., Milchberg, M. H., Rienstra, C. M. & Murphy, C. J. Biologically Representative Lipid-Coated Gold Nanoparticles and Phospholipid Vesicles for the Study of Alpha-Synuclein/Membrane Interactions. *ACS Nano* **17**, 20387–20401 (2023).

5. Middleton, E. R. & Rhoades, E. Effects of Curvature and Composition on α-Synuclein Binding to Lipid Vesicles. *Biophys J* **99**, 2279–2288 (2010).

6. Chandra, S., Chen, X., Rizo, J., Jahn, R. & Südhof, T. C. A Broken α-Helix in Folded α-Synuclein. *Journal of Biological Chemistry* **278**, 15313–15318 (2003).

7. Fredenburg, R. A. *et al.* The Impact of the E46K Mutation on the Properties of α-Synuclein in Its Monomeric and Oligomeric States. *Biochemistry* **46**, 7107–7118 (2007).

8. Kiskis, J., Horvath, I., Wittung-Stafshede, P. & Rocha, S. Unraveling amyloid formation paths of Parkinson's disease protein α-synuclein triggered by anionic vesicles. *Q Rev Biophys* **50**, e3 (2017).

9. Ladokhin, A. S., Fernández-Vidal, M. & White, S. H. CD Spectroscopy of Peptides and Proteins Bound to Large Unilamellar Vesicles. *J Membr Biol* **236**, 247–253 (2010).

10. Fusco, G. *et al.* Direct observation of the three regions in α-synuclein that determine its membrane-bound behaviour. *Nat Commun* **5**, 3827 (2014).

11. Klein-Seetharaman, J. *et al.* Differential dynamics in the G protein-coupled receptor rhodopsin revealed by solution NMR. *Proceedings of the National Academy of Sciences* **101**, 3409–3413 (2004).

12. Makasewicz, K. *et al.* Cooperativity of α-Synuclein Binding to Lipid Membranes. *ACS Chem Neurosci* **12**, 2099–2109 (2021).

13. Stephens, A. D. *et al.* α-Synuclein fibril and synaptic vesicle interactions lead to vesicle destruction and increased lipid-associated fibril uptake into iPSC-derived neurons. *Commun Biol* **6**, 526 (2023).

14.   Whittaker, H. T., Qui, Y., Bettencourt, C. & Houlden, H. Multiple system atrophy: genetic risks and alpha-synuclein mutations. *F1000Res* **6**, 2072 (2017).

CHAPTER 6

STRUCTURAL CHARACTERIZATION OF THE LIPID MEMBRANE-BOUND STATE

OF hTIM3

6.1. Introduction

The binding of the IgV domain of the human variant of TIM3 (hTIM3) to phosphatidylserine (PS) is known to play an important functional role within the immune response,[1] but little structural detail is known about the membrane-bound state of hTIM3. Recent simulations of hTIM3 have suggested that the protein may undergo a conformational switch upon associating with lipid membranes – a switch that did not occur for its murine counterpart, mTIM3.[2] Additionally, previous work within this field has suggested that hTIM3 has a lower affinity for PS in comparison to the murine variant (mTIM3).[3,4]

Within this chapter, we present how methodologies developed and used to determine the lipid membrane-bound state of mTIM3 were applied to the human variant. Detailed structural analysis of hTIM3 can provide key insight into the state of the IgV domain and the corresponding protein:lipid contacts that differentiate hTIM3 from mTIM3. With the field's increasing interest in the targeting of hTIM3, this structural detail can aid in therapeutic development for the treatment of immune exhaustion within prolonged stimulation during cancer and chronic viral infections.[5]

Our approach primarily entails structural characterization using a combined approach through MD and XR, as detailed in **Chapter 3.3.** However, in applying the protocols to the hTIM3 system, we encountered several unexpected complications, potentially linked to the differences between the murine and human TIM3 structures and bound states. The following sections discuss these complications and the rationalizations behind the adjustments to our approach to the hTIM3

106

system. Based on the available data, we end with our general conclusions on the structural features of the hTIM3 lipid-membrane bound state.

6.2. Structure of the IgV domain of TIM3 and the Ca²⁺ and PS-Binding Pocket

The structure of TIM3 has primarily been studied through X-ray crystallography, characterizing its non-lipid bound state.[6] The IgV domain of TIM3 is composed of two antiparallel β-sheets held by three disulfide bonds, forming the AGFCC'C" and BED faces **(Fig. 6.1a)**. Within this structure, the FG and CC' loop form a cleft, which coordinates with Ca^{2+} and the headgroup of PS. Both FG and CC' loops insert into the membrane and the BC loop makes peripheral contacts in the membrane-bound state for mTIM3.[7]

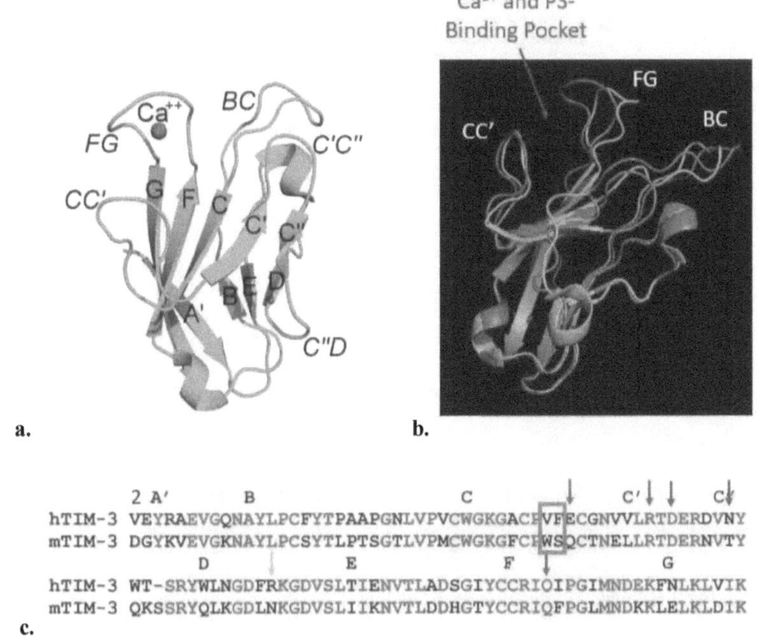

Fig. 6.1. a. Strand structure of hTIM3 **(PDB 6DHB)**. **B.** Backbone alignment of mTIM3 (purple, **PDB 3KAA**) and hTIM3 (cyan, **PDB 5F71**). **C.** Sequence alignment between hTIM3 and mTIM3

with conserved residues shown in red. V38 and F39 on the CC' loop are circled in green. E40, R47, and Q91 are identified by blue arrows. D49 and N54 are identified by purple arrows and R67 is identified by an orange arrow. Figures **a** and **c** adapted from Gandhi *et al.*[6]

While the human and murine variants of TIM3 share over 60% sequence similarity, resulting in similar IgV folds **(Fig. 6.1b,c)**, several key differences in the FG, CC', and BC loops may contribute to differences in membrane affinity. These differences include the following:

- **CC' loop:** The hydrophobic, aromatic tryptophan (W41) and polar serine (S42) residues in mTIM3 are replaced with the smaller valine (V38) and the hydrophobic, aromatic phenylalanine (F39). Within a membrane-bound state, these substitutions may result in an orientation tilt for hTIM3 to accommodate hydrophobic contacts between F39 and lipid tails. The neutral glutamine (Q43) in mTIM3 is replaced with the negatively charged glutamic acid (E40), which may alter the stability of the salt bonds formed with arginine (R50 and R47, respectively) on beta-strand C'. The stronger salt bond may as well contribute to rigidity of the pocket.

- **FG loop**: The sequence and structure of the FG loop is similar between mTIM3 and hTIM3, where the negatively charged residues D102 and D98, respectively, coordinate with the Ca^{2+} in the pocket. However, mTIM3's leucine (L99) is replaced with the bulkier isoleucine (I95) in hTIM3, which may increase the barrier for insertion for the FG loop.

- **BC loop**: The BC loop of hTIM3 contains one additional proline residue (P18, P24, P31 in mTIM3 and P15, P20, P23, and P28 in hTIM3), likely contributing to loop rigidity. Previous work by our group has found that differences in the BC loop between the HBA and Balb/c variants of mTIM3 contribute to the rigidity of the PS-binding pocket, which in turn results in a lower affinity for PS.[7] These previous results warrant exploration into the contributions of the BC loop to the membrane-bound state of hTIM3.

Upon binding to the membrane, the PS-binding pocket of mTIM3 opens to accommodate Ca^{2+} and PS, and except for the orientation of residues on the FG and CC' loops, the overall structure of mTIM3 largely remains the same. In contrast, Weber *et al.* found that hTIM3 undergoes a conformational switch when they performed MD simulations of the IgV domain with a Ca^{2+} and short-tailed PS in the pocket.[2] While the short-tailed PS does not represent the entirety of membrane contacts, the results suggest that pocket engagement and the corresponding lipid:pocket contacts in hTIM3 may differ from mTIM3.

This conformational switch is proposed to occur in two parts[2]: the introduced negative charge to the pocket from the PS headgroup allows for the formation of a salt bridge between residues E40 and R47 (**Fig. 6.2**). This releases N54, which was previously transiently bound to R47, to form contacts with D49. In turn, Y55 becomes solvent-exposed and can collapse into W61 through hydrophobic anchoring. To our knowledge, the membrane-bound state of hTIM3 has not yet been determined through either MD or experimentally. Therefore, it is important to characterize the contributions of additional membrane contacts and experimentally confirm this conformational switch.

To date, there are seven crystal structures of the IgV domain of hTIM3 that have been deposited to the Protein Data Bank, none of which were co-crystallized with a PS in the pocket.[6,8–11] While none of these cleanly represents a membrane-bound state, it is important to discuss differences in the published structures' strand arrangements and pocket engagements with other ligands to provide insight into conformational rearrangements that may occur upon PS-binding. For each of the published structures, we will compare the differences in the key residues identified above, which are also summarized in **Table 6.1,** while representatives of the various found states are shown in **Fig. 6.2**.

- **PDB 5F71**[8]: This structure is hTIM3 alone with nothing in the PS-binding pocket. Q91 is tilted away from the E40-R47 salt bridge, and the D49-N54 contact is not formed. Y55-W61 contact is formed.

- **PDB 6DHB**[6] **and 6TXZ**[9]: These two structures are very similar to one another, where **6DHB** was co-crystallized with a Ca^{2+} and a benzoate in the pocket and **6TXZ** was co-crystallized with a tyrosine side chain from an antibody inserted into the pocket. It is worth noting that the carboxyl group of benzoate and the hydroxyl on the tyrosine may mimic the effects of PS in the pocket by interacting similarly to its negatively charged serine headgroup. For both, the E40-R47 salt bridge is formed with Q91 extending towards the salt bridge, D49-N54 contact is formed, but Y55 is not collapsed into W61.

- **PDB 7KQL**[10]: This structure of hTIM3 was co-crystallized with an antibody, with nothing inserted into its PS-binding pocket. Q91 is tilted away from E40, and the E40-R47 salt bridge is replaced by the D49-R47 salt bridge. Similar to **5F71**, N54 does not form a contact with D49 and Y55 is collapsed into W61.

- **PDB 7M3Y/7M3Z/7M41**[11]: These three structures were all co-crystallized with small molecules that bind to the C"D loop and a Ca^{2+} in the PS-binding pocket. Overall, all three structures are very similar to one another and very similar to **6DHB**: Q91 extends towards the E40-R47 salt bridge, D49-N54 contact is formed, and Y55 is not collapsed into W61. It is worth noting that the small molecules form direct hydrophobic and π-π stacking contacts with W61 and may play a role in preventing the Y55-W61 contact from forming.

Overall, we can see that the seven published structures and their folds largely fall into two groups: one in which the pocket has no ligands (**5F71** and **7KQL**, **Fig. 6.2a**) and the other in which the pocket is engaged with Ca^{2+} or another ligand (**6DHB, 6TXZ, 7M3Y, 7M3Z,** and **7M41, Fig. 6.2b**). For the first group, Q91 tilts away from E40, D49-N54 contact does not form but Y55-W61 contact does. In contrast, the pocket engagement in Group 2 results in a Q91 tilting towards the E40-R47 salt bridge, D49-N54 contact forms, but Y55-W61 contact does not. These comparisons suggest that it is possible that pocket engagement by either Ca^{2+} or PS results in a shift in Q91 towards the E40-R47 salt bridge. It also appears that rearrangements to the D49-N54 and Y55-W61 contacts are correlated.

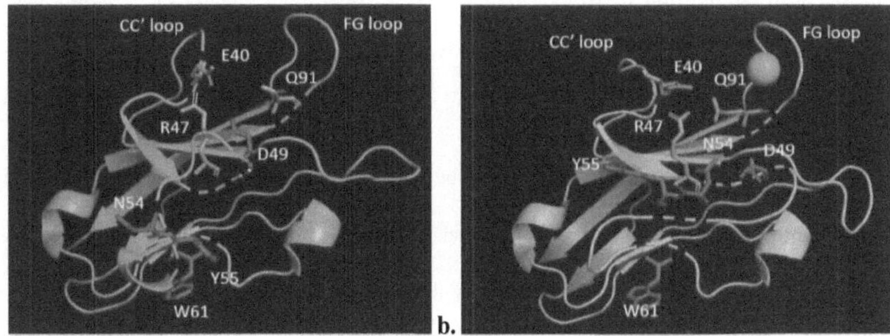

Fig. 6.2. Representative structures of the two general groups of published hTIM3 crystal structures. **a.** "Group 1", **PDB 5F71**[8]. **b.** "Group 2", **PDB 7M3Y**[11]. E40-R47-Q91 bridges are shown in orange, Y55-W61 is shown in purple, D49-N54 is shown in red. Note the rearrangement of strands that leads to differences in the formation of the D49-N54 and Y55-W61 contacts.

Table 6.1. Summary of structural features of the existing crystal structures. The two general structural groups described above are highlighted in blue and orange.

PDB	Co-crystallization Factors	E40-R47 contact?	E40-R47-Q91 triad?	D49-N54 contact?	Y55-W61 contact?
5F71	None	Yes	No	No	Yes
6DHB	Ca^{2+}, Benzoate	Yes	Yes	Yes	No

Table 6.1, continued.

6TXZ	Antibody, tyrosine in pocket	Yes	Yes	Yes	No
7KQL	Antibody	No	No	No	Yes
7M3Y/7M3Z/7M41	Ca^{2+}, small molecules	Yes	Yes	Yes	No

Outside of the crystal structures, where the process of crystallization may be packing the protein into non-physiologically-relevant states, there is limited structural characterization of hTIM3 either in solution or at the membrane. However, it is important to note that Gandhi et al.[6] observed minimal chemical shift in the ^{1}H-^{15}N HSQC spectra of hTIM3 upon addition of Ca^{2+} for residues E40 and R46, suggesting that there may not be a discernable change in the contacts of these residues in the presence and absence of Ca^{2+}. It is unclear from these data which group of crystal structures better reflects a true solution state.

Additionally, these structural features differ between the crystal structures and those found from the MD simulations performed by Weber et al., where the Y55-W61 contact is formed only after engagement of the pocket.[2] It is not clear if these differences arise from experimental artifacts (such as crystal packing or starting state for MD simulations), but they highlight the need for a combined experimental and MD approach to resolve the membrane-bound state of hTIM3. The following sections aim to resolve these discrepancies by taking a closer look into how membrane-engagement affects global protein structure. We first discuss the experimentally-available binding assays to determine the Ca^{2+} and PS specificity of the IgV domain of hTIM3 (**Chapter 6.3**) and then explore the structural features of the membrane-bound states through MD and XR (**Chapters 6.4-6.5**).

6.3. Protein:Membrane Binding for hTIM3

6.3.1. Tryptophan Fluorescence

For the murine TIM variants, we use a tryptophan fluorescence assay to quantify the Ca^{2+} and PS dependence of protein:membrane interactions,[7] but in comparison to the other variants, hTIM3 does not have a tryptophan residue placed on the FG or CC' loops that would insert into the membrane. However, the proposed conformational switches described above may result in a change to the emission spectrum of W61 as it changes its association state with Y55. When the Y55-W61 contact is formed, the tryptophan may be more shielded from the aqueous environment, possibly within the detection range of tryptophan fluorescence.

However, our experimental data show no change in the tryptophan emission spectrum of hTIM3 upon the addition of PS-containing lipid membrane or Ca^{2+} (**Fig. 6.3**), eliminating tryptophan fluorescence spectroscopy as an option to quantify the protein's affinity for PS and Ca^{2+}. The lack of visible changes in the tryptophan fluorescence spectrum suggests that any other structural changes or conformationally changes of hTIM3 only minimally alter the environment of exposed tryptophan residues.

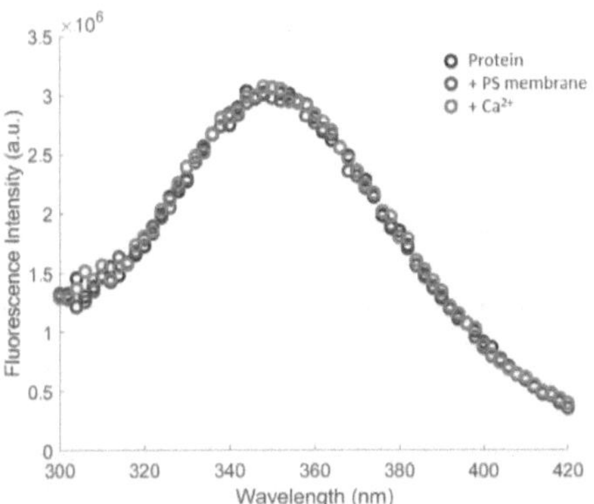

Fig. 6.3. Representative tryptophan fluorescence emission spectra for 170 nM hTIM3 in the presence of 300 μM 70:30 POPC:PS lipid vesicles with and without Ca^{2+}.

6.3.2. Sucrose Sedimentation Assay

As tryptophan fluorescence is unable to verify PS-association of hTIM3, we use a sucrose sedimentation assay to test the Ca^{2+} and PS dependence of hTIM3's membrane binding. In short, a sucrose sedimentation assay identifies any membrane-associated states and can be used to explore the conditions under which a protein would bind to a vesicle by using "heavy" liposomes, loaded with sucrose, to isolate bound protein from unbound[12] (full description of the methodology in **Chapter 3.2.1**).

To test the binding behavior of hTIM3, we incubated protein with sucrose-loaded LUVs composed of POPC (**Fig. 6.4a**) and 7:3 POPC:POPS (**Fig. 6.4b**) under varying calcium concentrations (0, 0.3, 0.5, and 2 mM Ca^{2+}). As shown in **Fig. 6.4a**, hTIM3 does not appear to bind to POPC vesicles under any calcium concentrations, as evident by the lack of a band corresponding to its molecular weight in the "bound" lanes (2, 4, 6, 8, and 10) and the presence of a strong band in the "unbound" lanes (3, 5, 7, and 9) similar to the protein-only control (lane 1). A similar experiment with vesicles composed of 100% DOPC shows similar behavior, where hTIM3 does not appear to bind to the LUVs under any calcium concentrations.

However, hTIM3 binds in a Ca^{2+}-dependent manner to vesicles composed of 7:3 POPC:POPS. In comparison to the 0 mM Ca^{2+} sample (lanes 3 and 4), we see increasing hTIM3 binding with increased calcium concentrations (lanes 5-10), where the intensity of the "bound" fractions relative to the "unbound" increases across each pair of lanes with higher calcium concentrations. Similar binding behavior was observed for vesicles composed of 7:3 DOPC:DOPS. Altogether, these results suggest that hTIM3 binds to lipid membranes in a PS-specific manner that, in turn, is calcium concentration-dependent. While this assay does not provide precise quantitative information for the affinity of hTIM3 for Ca^{2+} or PS, it does indicate that the expressed protein is not simply adsorbing but is binding to the lipid membrane in a lipid-specific manner.

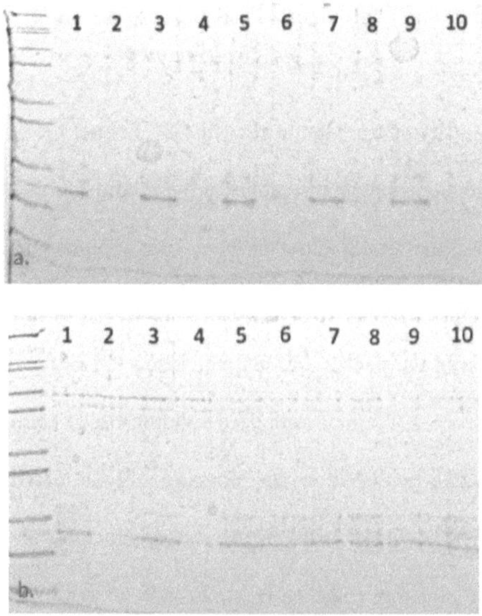

Fig. 6.4. Gel-electrophoresis results of a lipid sedimentation assay for hTIM3 binding to **a.** POPC and **b.** 7:3 POPC:POPS vesicles. Leftmost lane for both gels is the Mark12 standard. For the pair of lanes of each lipid and calcium condition, the "unbound" fraction, corresponding to the supernatant, is on the left and the "bound" fraction, corresponding to the vesicle pellet, is on the right. **Lanes 1-2**: 0 mM lipid, 2 mM Ca^{2+}, **lanes 3-4:** ~2.5 mM lipid, 0 mM Ca^{2+}, **lanes 5-6:** ~2.5 mM lipid, 0.3 mM Ca^{2+}, **lanes 7-8:** ~2.5 mM lipid, 0.5 mM Ca^{2+}, and **lanes 9-10:** ~2.5 mM lipid, 2 mM Ca^{2+}.

6.4. MD Simulations of hTIM3

MD simulations of hTIM3 can provide insight into the membrane-bound state of the protein. As discussed in **Chapter 6.2**, the protein's global state appears to depend on the engagement of the PS-binding pocket, but published results do not currently extend to include the other lipid:protein contacts that occur at the membrane. The following section details the results of our MD simulations of hTIM3 that explore the stability and binding-behavior of the various proposed structures of the protein. Within our analysis, we emphasize the membrane-orientation of the protein along with a comparison of how conformational states relate to formed protein:lipid contacts. This section additionally provides a basis for comparison for structural analysis of XR data in **Chapter 6.5** below.

6.4.1. Initialization: Stability of Ca^{2+} ion in the Binding Pocket

The initialization of simulations that would capture protein:lipid membrane interactions requires a suitable structure of hTIM3 that would allow for the protein to bind to the membrane from a stable solution-based starting state. To generate this structure, we attempted to equilibrate several hTIM3 structures with a Ca^{2+} ion in the pocket, but in comparison to simulations of mTIM3, we found this to be a non-trivial process. We first initialized solution-based equilibrations of hTIM3 with Ca^{2+} in the pocket from two published crystal structures, PDBs: **6DHB** and **7M3Y** (arbitrarily chosen from the **7M3Y/7M3Z/7M41** series, as these three structures all have near perfect pocket and backbone alignment[11]). Despite the co-crystallization of both structures with Ca^{2+} in the pocket, the solution equilibrations of these starting states resulted in the disassociation of the Ca^{2+} ion after ~10-20 ns of simulation across all 5 trials of each structure.

These results point to the instability of the Ca^{2+}-coordination of these published states, which may be attributed to the other co-crystallization factors of the two structures. The presence of the benzoate in the pocket of PDB **6DHB** donates electron density to the pocket, potentially stabilizing the E40-R47-Q91 bridge that tilts away from the pocket (**Fig. 6.5**). Meanwhile, PDB **7M3Y** is co-crystallized with the small molecule YQ7 that binds to the C"D strand, possibly preventing strand rearrangement and stabilizing the E40-R47-Q91 bridge. Further discussion of these states and their relevance to stability of the Ca^{2+}-bound state can be found below in **Chapter 6.4.3**, but within these solution simulations, the removal of the small molecule and the benzoate likely contributes to the instability of Ca^{2+}-coordination within their respective simulations.

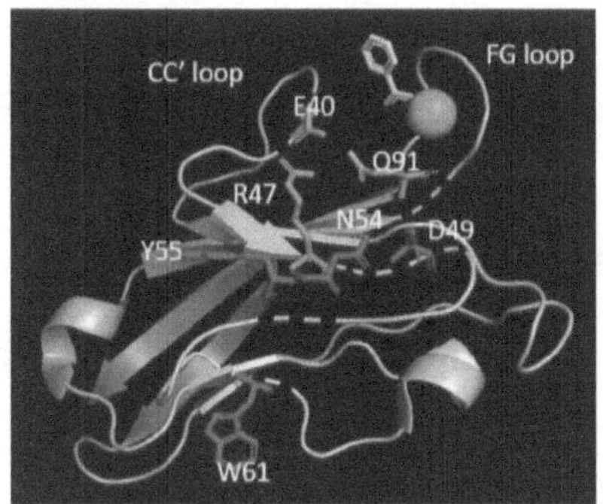
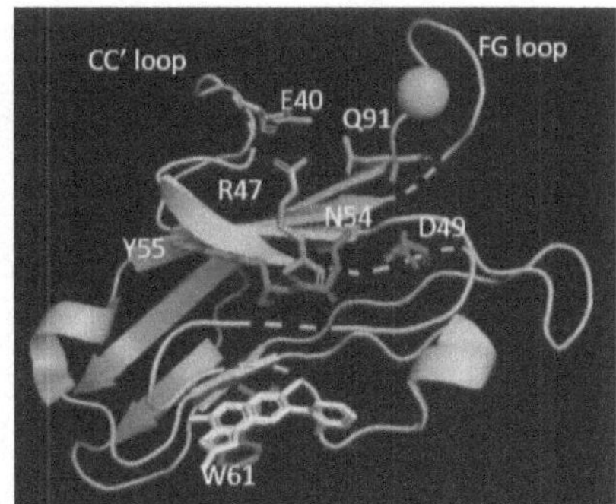

Fig. 6.5. Crystal structures of PDB **a. 6DHB** (light purple) with benzoate (yellow) bound, and **b. 7M3Y** (cyan) bound to small molecule YQ7 (yellow), in the presence of Ca^{2+} (green). Both structures show the stable E40-R47-Q91 bridge (orange) and lack of Y55-W61 contact (magenta).

As neither of these structures appear to be true representatives of a Ca^{2+}-only bound states, we turned to structure PDB **5F71**, a crystal structure of hTIM3 with an empty pocket and a different backbone configuration (**Chapter 6.2**). However, placing a Ca^{2+} ion into the pocket based on homology modeling to mTIM3 was unsuccessful, as the Ca^{2+} ion immediately dissociated from the protein across all 5 trials. Overall structural similarities within the general groups of the published crystal structures of hTIM3 (**Chapter 6.2**) suggest that initialization of simulations from the other crystal structures would likely be similarly unsuccessful. This contrasts with simulations of the other murine[7] members of the TIM family where stable Ca^{2+}-bound solution states were easily equilibrated from their respective crystal structures. These results suggest that it is possible that a Ca^{2+}-only bound state of hTIM3 is not amenable to crystallization or may only exist as a transition state between an empty pocket and a PS-engaged state. Further analysis of membrane simulations in **Chapters 6.4.2-6.4.3** below expands upon the factors that may contribute to these results.

Regardless of these results, we still needed a stable solution state structure of hTIM3 for membrane simulations. We surmised that a Ca^{2+}-equilibrated state may only be achievable in the presence of an additional ligand in the pocket that mimics the headgroup of a PS molecule. For this, our collaborator Jeff Weber from the International Business Machines (IBM), extended the solution equilibration of hTIM3 with a Ca^{2+} ion and the short-tailed PS molecule in the pocket from the study detailed above[2] (**Chapter 6.2**). For this structure, we additionally appended residues at the N-terminus to match the protein construct used within the experimental work in **Chapters 6.3** and **6.5**.

With this structure, we saw a stable Ca^{2+}- and PS-bound state over the course of 20 μs of simulation time. A few key features of this structure differ from the general crystal structures for hTIM3 and Ca^{2+} and PS-engaged mTIM3 (**Fig. 6.6a**). Most noticeable, the FG loop twists and splays out, with the Ca^{2+} ion sitting more loosely within the pocket. We observe that this state does not contain the intact E40-R47-Q91 bridge seen in some of the pocket-engaged crystal structures, but instead E40 and Q91 extend towards the Ca^{2+} ion, while R47 tilts away from this contact (**Fig. 6.6b**), resulting in rearrangement of the C''D strand that allows for the Y55-W61 contact to form. Interestingly, these contacts are similar to the ones observed in the hTIM3 crystal structures that do not have pocket engagement (PDB **5F71** and **7KQL**).

This state is referred to as "State A," which we use as a starting state for the initialization of membrane simulations after removing the PS molecule from the pocket. We reason that this state contains pocket and backbone configurations that are better suited for both Ca^{2+}-stability and for the insertion of a PS headgroup into the pocket. While simulations from this state will not reflect full conformational changes that a protein would undergo, they can identify membrane-bound states and the protein:lipid contacts that contribute to stability of the protein at longer timeframes.

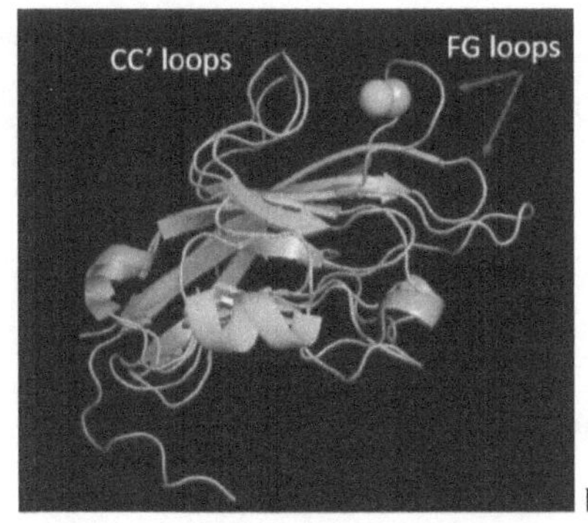

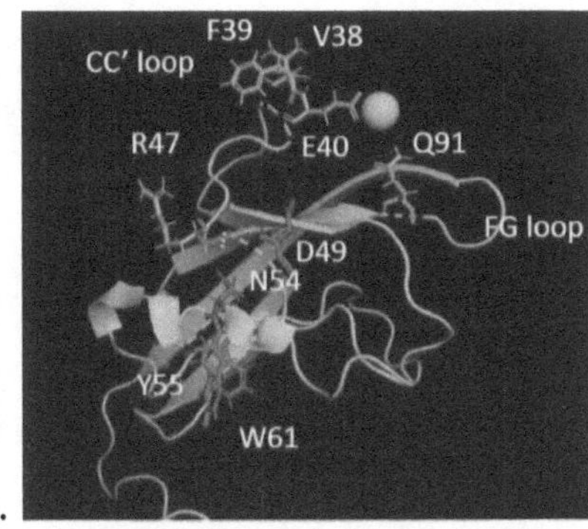

Fig. 6.6. "State A" of hTIM3 (cyan) found from extended simulations of hTIM3 with Ca^{2+} and PSF in the pocket. **a.** Overlay of backbone of hTIM3 with mTIM3 (PDB **3KAA**, pink) with Ca^{2+} ions shown in their respective color. **b.** Detailed residue arrangement. Of note, E40-Q91 extend towards the Ca^{2+}, while R47 tilts away from the bridge due to strand rearrangement, which allows Y55-W61 contact to form.

6.4.2. HMMM Simulations of hTIM3 from State A

While State A mimics the pocket engagement of a PS-bound state, it does not include the additional lipid:protein contacts that are formed at the membrane. In our aim to identify these along with the protein's orientation at the membrane, we turned to HMMM simulations (**Chapter 3.3.4.2**). For this, we initialized 10 trials of hTIM3 in State A with a HMMM membrane composed of 7:3 PC:PS with the protein placed in solution to mimic a binding event.

As expected, some of these trials resulted in protein diffusing away from the membrane or binding in an upside-down orientation (e.g., with N-terminus inserting into the membrane) and were unfit for extension, but all retained a stably-bound Ca^{2+} ion. A set of these trials showed promising binding events with the protein engaging with PS on the membrane. Over the course of 200 ns of simulation time, consistent binding features across this subset of trials include the following (**Fig. 6.7**):

- The hydrophobic V38 and F39 residues on the CC' loop orient to insert into the membrane, forming contacts with the lipid tails of POPC and POPS lipids.

- Y55-W61 contact remains stable.

- E40 remains stably associated with the Ca^{2+} ion, while Q91 forms contacts with POPS lipids.

- K100 forms stable contacts with POPS lipids.

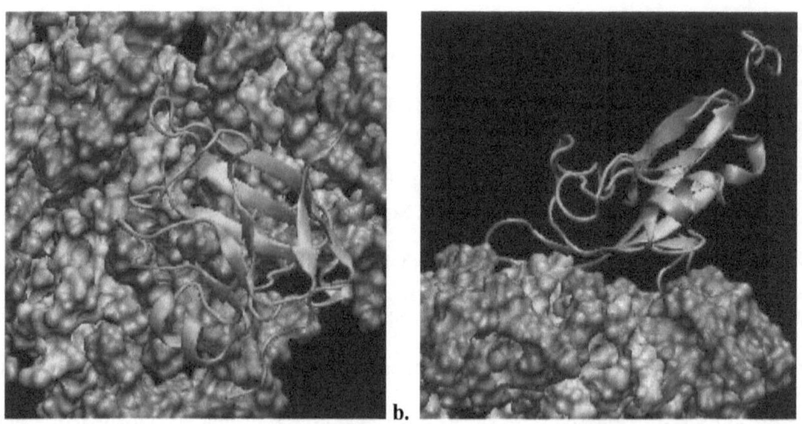

Fig. 6.7. Representative frame of hTIM3 (cyan) following 200 ns simulation of State A.

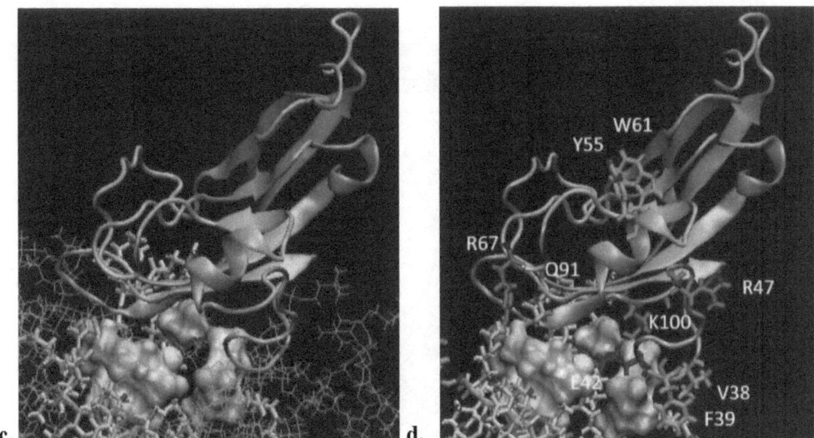

Fig. 6.7, continued. "State B" representative frame of hTIM3 (cyan) following 200 ns simulation of State A on an HMMM membrane. Panels **a** and **b** show vertical and horizontal perspectives of the protein on the 7:3 PC (tan): PS (pink) membrane. Panel **c** visualizes protein from a ~30° rotation from the membrane plane. PS headgroups within 5 Å of the protein are visualized in surface representation. Panel **d** identifies key residues discussed within the text. Positively-charged, negatively-charged, and polar residues are shown in red, blue, and tan, respectively. Nonpolar residues are shown in green.

While 200 ns simulations are capable of identifying these contacts, the timescale is too short to observe any strand rearrangements or the binding event's effects on secondary structure, as these occur on the µs timescale. To investigate whether the global protein structure and the conformational switch are retained on the HMMM membrane, we chose the final frame of one of these trials, which we refer to as "State B," to spawn trajectories that were extended into the µs timescale. The extension of these HMMM trials was performed by our collaborator Jeff Weber, and a representative image of the final state from his work, "State C," is shown in **Fig. 6.8**.

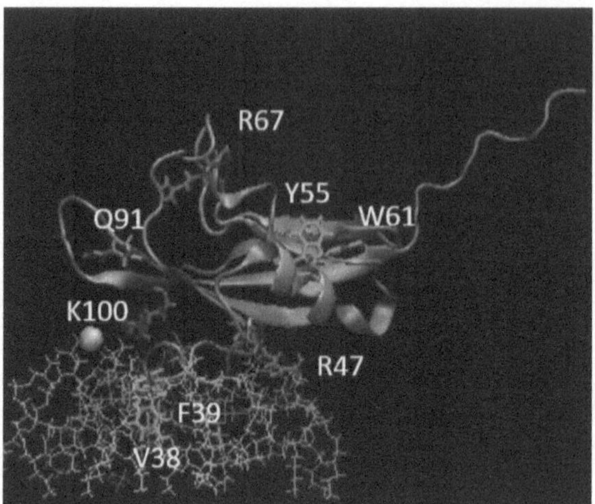

Fig. 6.8. Horizontal view of "State C" representative frame of hTIM3 (cyan) following 1 μs simulation of State A on a 7:3 PC:PS HMMM membrane (horizontal orientation of protein can be compared to **Fig. 6.7b**). Positively-charged, negatively-charged, and polar residues are shown in red, blue, and tan, respectively. Nonpolar residues are shown in green. For visualization purposes, PC lipids are not shown.

Over the full 1 μs, hTIM3 retained stable contacts with the Ca^{2+} ion and the PS headgroups as the protein gradually tilted horizontally, resulting in contacts between R47 and a PS in the membrane (straight-on perspectives of the bound protein are shown in **Fig. 6.9a**). Overall, the protein does not appear to undergo any large conformational changes on either the ns or μs timescales with good alignment of the protein's backbone across the simulation (**Fig. 6.9b**), but intriguingly, the FG loop continues to gradually twist outward from the binding pocket. Of note, the protein appears to engage with multiple PS headgroups in a halo-like orientation around the pocket (**Fig. 6.7c**) with two PS headgroups forming contacts with K101 and two to three PS headgroups on the other side of the Ca^{2+} ion. In comparison to the murine TIMs, none of these PS molecules directly insert into the pocket, and instead appear to cluster around the pocket, partially stabilized by the Ca^{2+} ion and positively-charged or polar residues.

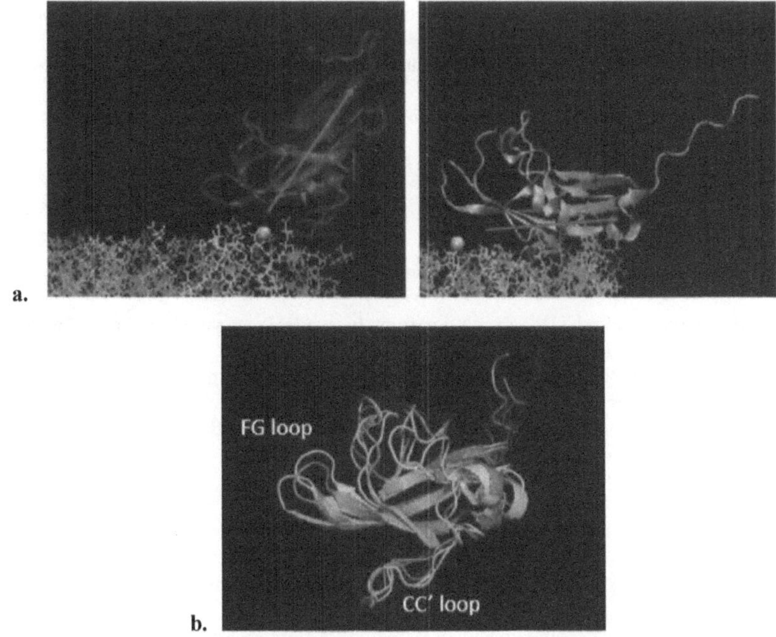

Fig. 6.9. a. Alignment of State B (left) and State C (right) with respect to the membrane to show the progressive hoizontal tilt from 200 ns (purple) to 1 μs (cyan). Red arrows are included to guide the eye for general protein orientation. **b.** Backbone alignment of hTIM3 from starting State A (purple), State B (cyan) after ~200 ns of simulation, and State C (green) following ~1 μs of simulation.

Altogether our HMMM simulations successfully capture the binding of hTIM3 to PS headgroups within the membrane. The extensions of these trials show a stable secondary structure and backbone arrangement of the protein, where the Y55-W61 contact is formed but E40-R47-Q91 is not. We note that in comparison to mTIM3, the protein appears to favor a horizontal orientation, potentially mediated by a contact between residue R47 and the negatively charged PS. The apparent cooperativity between the PS contacts with Ca^{2+} and residue R47, similar to R50 in mTIM3[7], likely increase hTIM3's sensitivity to PS surface density. The competing E40-R47 contact in hTIM3, which does not exist in the murine protein, may explain the reported lower affinity of hTIM3 for lipid membranes[3] in comparison to mTIM3[13].

6.4.3. Membrane Docking Simulations of hTIM3: Stability of Protein Orientation

While membrane simulations starting from State A allow us to capture a binding event and follow the stability of the membrane-bound structure, the starting state potentially bypasses physiologically-relevant global conformational changes. As we are unable to initialize a stable Ca^{2+}-only bound state for hTIM3 in solution from the available crystal structures (**Chapter 6.4.1**), we turned to membrane-docked simulations in the hope that PS at the membrane would stabilize the Ca^{2+} in the pocket and allow us to track conformational changes throughout a membrane-bound state of protein structures that were unstable with just the Ca^{2+} ion in the pocket. With membrane docking, we additionally aimed to determine if the horizontal orientation found from extending State A is reproducible and representative on re-initialized membranes which would likely contain a different distribution of lipids and would require re-equilibration of the protein to form its preferred lipid:protein contacts.

6.4.3.1. Membrane Docking: Stability of Horizontal Orientation for State C

To first test the representative orientation of State C, we initialized a set of 10 trials of State C in a horizontal orientation or in a vertical orientation with respect to the membrane (5 trials for each). This horizontal orientation was modeled after the orientation found in the extended membrane simulation from State C, where the PS used to dock in the membrane was chosen to be the one that most closely coordinated with the Ca^{2+} ion. For the vertical orientation, the placement of the Ca^{2+} and PS in the pocket was modeled based on backbone alignment with a representative membrane-bound state of mTIM3.[7] The orientations and PS alignment for both orientations are shown in **Fig. 6.10**.

 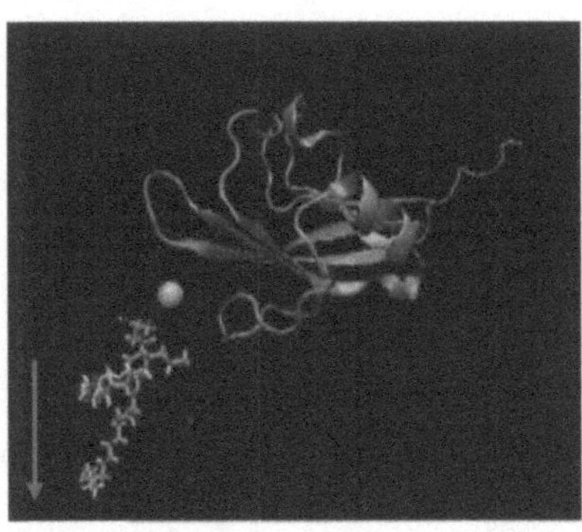

Fig. 6.10. Initialized vertical (left) and horizontal (right) orientations for hTIM3 in State C for docked trials. Membrane z-direction is shown by red arrow. Ca^{2+} ion and POPS are shown in yellow and pink, respectively. Initialization of docked simulations with **7M3Y** used the same orientations with backbone alignment of **7M3Y** to State C.

We then extended each of these 10 trials to 200 ns, with the results summarized in **Table 6.2**, where we observed several trends among the trials. State C had a clear preference for initialization in the horizontal orientation with 4/5 trials retaining their Ca^{2+} and PS-bound states in the horizontal orientation. In comparison, initialization in the vertical orientation resulted in the loss of PS-specificity in the binding in 3/5 trials with the protein either completely dissociating from the membrane or tumbling into a non-physiologically relevant state (e.g., upside down).

Table 6.2. Summary of hTIM3 membrane docking simulations for State C. A trial was determined to be "stable" if it retained Ca^{2+} in the pocket and a reasonable binding orientation (e.g., not upside down). Contacts were determined if within 5 Å of each other. Number of PS engaged with pocket was identified as number of PS headgroups within 5 Å of calcium or residue K100.

Trial	Stable?	R47 coordinated with PS?	Y55-Y61 contact?	E40-Ca^{2+} contact?	Number of PS engaged with pocket
State C, Vertical	Yes	No	Yes	Yes	3

Table 6.2, continued.

State C, Vertical	Yes	No	Yes	Yes	1
State C, Vertical	No				
State C, Vertical	No				
State C, Vertical	No				
State C, Horizontal	No				
State C, Horizontal	Yes	Yes	Yes	Yes	1
State C, Horizontal	Yes	Yes	Yes	Yes	0
State C, Horizontal	Yes	Yes	Yes	Yes	1
State C, Horizontal	Yes	Yes	Yes	Yes	3

Interestingly, we observe that docking in hTIM3 in its horizontal orientation resulted in all 4 of the successful trial eventually forming contacts between R47 and a PS headgroup, further supporting the residue's role in stabilizing the protein on the surface. As this contact cannot immediately form in the vertical orientation, its absence may explain why the initialization of State C in a vertical orientation, where R47 cannot quickly form contacts with the PS without the protein first tilting, resulted in non-stable binding events.

Altogether this set of 10 trials suggests that the horizontal orientation of hTIM3 is stable, partially contributed to by the R47-PS contact. However, even with this data set, it is still not clear if a vertical binding event could eventually lead to a horizontal orientation as this tilt may happen on the µs timescale (as evident by the simulations of State A in **Chapter 6.4.2**). Extension of the stable states to µs could help elucidate the specific contacts that may contribute to the tilt.

6.4.3.2. Membrane Docking: Capturing Conformational Change of 7M3Y

In working with State C, we were concerned that the process by which we arrived at the structure for its precursor, State A, may prevent our simulations from observing physiologically-relevant conformational changes that would occur to experimentally-determined structures of hTIM3 on the lipid membrane. To address this concern, we chose PDB **7M3Y** as a starting state for our membrane docking simulations, arbitrarily chosen from the members of "Group 2" of the published crystal structures (**Chapter 6.2**). In choosing a structure from this group, we aimed to connect the differing strand arrangement of this set of structures with the conformational changes found through MD. Of note, the Ca^{2+}-coordination was not stable in our solution-based simulations of **7M3Y**, and choosing **7M3Y** had the additional benefit of observing how PS-coordination contributes to the ion's stability.

Two sets of 5 trails were initialized using backbone alignment with State C for both the horizontal and vertical orientations (**Fig. 6.10**) and extended to 200 ns. A summary of these results is displayed in **Table 6.3**.

Table 6.3. Summary of hTIM3 membrane docking simulations from PDB **7M3Y**. A trial was determined to be "stable" if it retained Ca^{2+} in the pocket and arrived at a reasonable binding orientation (e.g., not upside down). Contacts were determined if within 5 Å of each other. Number of PS engaged with pocket was identified as number of PS headgroups within 5 Å of calcium or residue K100.

Trial	Stable?	R47 coordinated with PS?	Y55-Y61 contact?	E40-Ca^{2+} contact?	Number of PS engaged with pocket
7M3Y, Vertical	Yes	Yes	No	No	2
7M3Y, Vertical	Yes	No	No	No	1
7M3Y, Vertical	Yes	No	No	No	4
7M3Y, Vertical	Yes	Yes	No	No	2
7M3Y, Vertical	Yes	Yes	No	No	4
7M3Y, Horizontal	No				
7M3Y, Horizontal	Yes	Yes	No	No	2
7M3Y, Horizontal	No				
7M3Y, Horizontal	No				
7M3Y, Horizontal	No				

Surprisingly, we found that membrane docking of **7M3Y** in the vertical orientation was stable, with all 5 trials retaining both Ca^{2+} and PS-coordination, but lost its Ca^{2+}-coordination in 4/5 trials of the horizontal orientation. This directly contrasts with the results found for State C, suggesting that the differences in the initial placements of the Ca^{2+} ion between the vertical and horizontal orientations likely contribute to the stability of the trials. As shown in **Fig. 6.10**, the Ca^{2+} ion in the horizontal orientation is pushed further outside the pocket and lies closer to the CC' loop, while in the vertical orientation, the Ca^{2+} is more closely associated with the FG loop. However, despite the differences in the starting placement among orientations, the E40-Ca^{2+} contact is formed in all State C trials but not in the 7M3Y trials, suggesting that backbone rearrangement may be needed for this contact to form. Without this rearrangement, it is likely that the Ca^{2+}-pocket distance may be too large, allowing for PS headgroups to strip the Ca^{2+} ion from **7M3Y** when initialized in the horizontal orientation.

We observed a trend among the 7M3Y trials docked in the vertical orientation. For the trials where R47 formed contacts with PS (trials 1, 4, and 5), the FG loop shifted and slightly twisted outward, potentially reflecting a transition state towards a structure similar to State C (**Fig. 6.11a,b**). In these trials, in comparison to trials 2 and 3, all observed a more horizontal bound state at the end of the 200 ns, this shift may be attributed to the freeing of the FG loop from the membrane. Out of these three, trial 5 had the most shifted FG loop and was paired with the largest rearrangement of strand C'C" (**Fig. 6.11c**), which, similar to the FG loop, appears to be in a transition state between the starting structure of 7M3Y and State C. This is additionally reflected in the rotation of Y55 (**Fig. 6.11d**). In comparison, trials 2 and 3 had near perfect alignment of the FG loops and residue Y55 with their starting state.

Across the trials for both State C and **7M3Y**, residues R47 and R67, located on opposite sides of the PS-binding cleft formed by FG and CC', appear to have a competing effect on the protein's orientation through their contacts with PS headgroups. Engagement of R47 with PS favors a tilted/horizontal state, while R67 appears to favor a more upright/vertical state. Within the 7M3Y, vertical trials, we observed that engagement with R47 would result in a gradual tilt in the protein – similar to the observed tilt in the transition from State A to State C.

Altogether these trials suggest that engagement of R47 with PS headgroups may precede the twisting of the FG loops, which is paired with larger strand rearrangements. While 200 ns is likely too short of a simulation time to observe these larger strand rearrangements in their entirety, the stability of the membrane-bound 7M3Y structure in its initial vertical orientation across all five trials along with the pairing of gradual tilt with strand shifts, is promising for these simulations to reveal a physiologically relevant transition. Extension of the 7M3Y trials to longer timescales may further identify the full collapse and provide sufficient simulation time to observe these shifts across all trials.

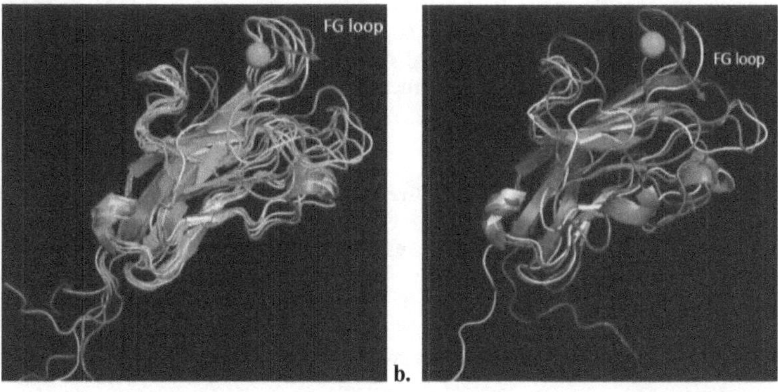

Fig. 6.11. Structural features of 7M3Y after 200 ns of simulation time.

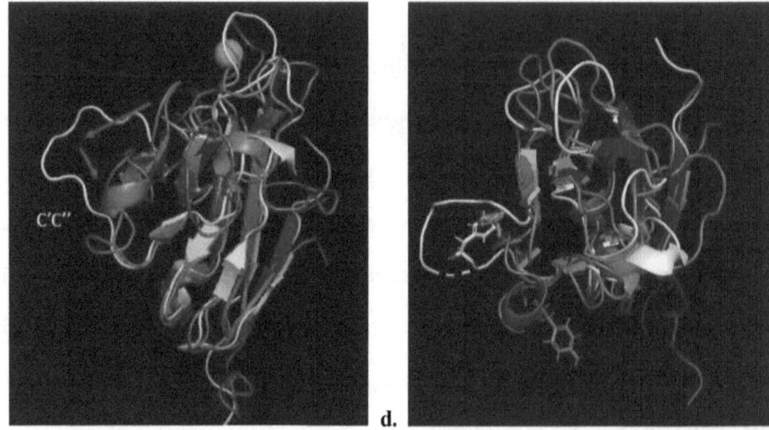

c.   d.

Fig. 6.11, continued. Structural features of 7M3Y after 200 ns of simulation time with docking on a 7:3 PC:PS HMMM membrane in the vertical orientation. All structures were aligned using backbone alignment to the initial crystal structure. Panel **a** shows shift of FG loop in trials with E46:PS contact with initial structure shown in pink, and trials 1-5 shown, respectively, in green, cyan, yellow, orange, gray. Panels **b, c,** and **d** show how trial 5 (gray) lies between initial **7M3Y** structure (pink) and State C (blue) with structural detail of C'C" and Y55 shown in **c** and **d**, respectively.

6.4.4. Conclusions from MD and Future Directions

MD simulations of hTIM3 reveal a few key insights surrounding structural features of its membrane-binding that differentiate it from its murine counterpart. First and foremost, extended simulations point towards stabilization of the protein in a horizontal orientation with respect to the membrane with the R47:PS contact appearing to play an important role within this orientation. This contact is additionally correlated with the absence of the E40-R47-Q91 bridge, and the difference between the human and murine proteins may thus be potentially related to the Q40E substitution in the human variant.

The membrane-bound state additionally contains strand arrangements that reflect those found in the solution stable state of hTIM3 with Ca^{2+} and PS in the pocket (State A), suggesting that this state may be representative of the true global conformational shifts that occur upon the protein engaging with PS-containing membranes. This state contrasts with the existing crystal structures, primarily through a tilted and splayed FG loop and the rearrangement of the C'C" strand that allows for the collapse of Y55 into W61. Intriguingly, some of these features, such as the Y55-W61 contact are only reflected in the crystal structures of hTIM3 with a non-engaged pocket (PDB 5F71, 7KQL) rather than an engaged one, as would be expected. While we have preliminary data suggesting that the protein may undergo a conformational change that would connect the crystal structure to State C, we have yet to observe the full transition – likely due to the limited simulation time.

It is additionally important to note that MD simulations have not yet extended to examine how membrane-binding would alter the structure of members from "Group 1" (**Chapter 6.2**). As these structures represent another global conformational state of the protein with experimental backing, it would be necessary to conduct similar docking trials from one of these members, like PDB **5F71**, to control for crystal packing and artificial strand arrangement within the crystallization process. Without these trials, we cannot definitively detail the protein's conformational change on the membrane. With this in mind, our preliminary results of the **7M3Y** structure indicate that hTIM3 may initially find the membrane in a vertical orientation and then equilibrate to a horizontal one once the required membrane-contacts are formed that allow for backbone rearrangement.

In the following section we aim to connect the observed structural features of hTIM3 in the MD simulations with experimental verification through XR with a focus on protein orientation when bound to a membrane.

6.5. X-ray Reflectivity of hTIM3

X-ray Reflectivity has been previously used to resolve the membrane-bound orientations of the murine TIMs[7,13]. We can apply a similar process for experimental analysis of the structure of the membrane-bound state of hTIM3 as experimental validation for the horizontal orientation proposed in **Chapter 6.4**. However, in performing XR experiments, we found hTIM3 to be a difficult protein to work with due to its low affinity for PS and susceptibility to aggregation. Here we detail the optimization of experimental conditions and protocol for acquiring XR scans of hTIM3, followed by an analysis and comparison of the collected data to the previously discussed structures (**Chapter 6.2** and **Chapter 6.4**).

6.5.1. Collection of XR Data for hTIM3

In the past, XR data of the murine TIMs were generally collected under the same general conditions: protein is injected under a lipid monolayer compressed to 25-30 mN/m in HBS buffer (10 mM HEPES, 150 mM NaCl, pH 7.2), followed by the addition of Ca^{2+}. These conditions resulted in reproducible data for the murine TIMs across various lipid and proteins samples, which were mostly stable to minor surface disruption (e.g., settled dust when sample chamber was opened for Ca^{2+} addition), and were stable for compression/expansion of the monolayer. However, our first attempt to collect hTIM3 data on a 7:3 SOPC:PS monolayer at 25 mN/m, in HBS buffer with 4 mM Ca^{2+} showed minimal binding, as evident by a lack of expansion of the surface area while the surface pressure was kept constant during the experiment, and little change in the measured XR curve from the lipid-only scan (**Fig. 6.12**).

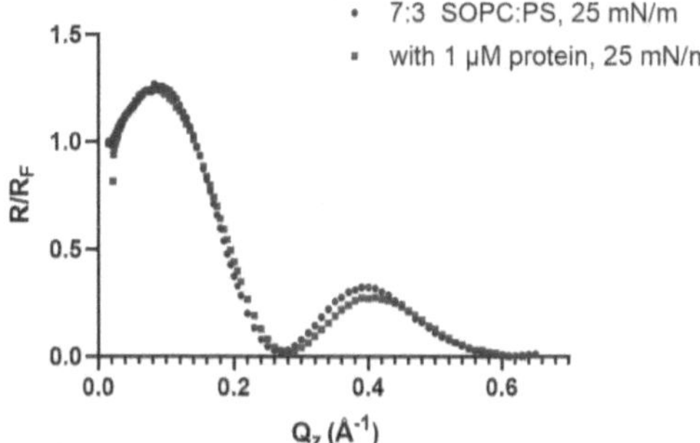

Fig. 6.12. XR curve showing minimal binding of hTIM3 to a 7:3 SOPC:PS monolayer at a surface pressure of 25 mN/m. Data were collected at 4 mM Ca^{2+} and in HBS buffer. Data are plotted as Fresnel Normalized Intensity (**R/R_F**) against the momentum transfer along the z direction (**Q_z**).

We surmised that the apparent lack of binding may be a result of hTIM3's low affinity for PS and attempted to expand the monolayer to the lower surface pressure of 23 mN/m to encourage protein insertion by decreasing the packing density of the lipid film and thus reducing the barrier to insertion. However, an initial area expansion was followed by a second rapid "runaway" expansion (**Fig. 6.13a**), where under the constant pressure control, we saw a continued increase in surface area, indicative of continuous protein insertion and a broadening of the first peak within the XR curve (**Fig. 6.13b**).

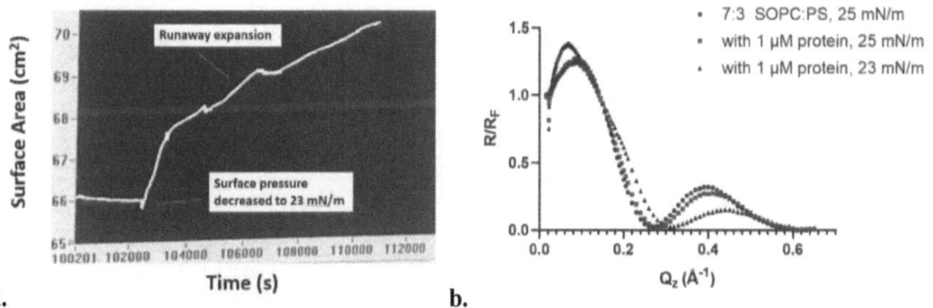

a.   b.

Fig. 6.13. hTIM3 XR data.

Furthermore, electron density fitting of the XR curve placed the protein's electron density at the same z position as the monolayer. All these indicate that the protein was not binding in a PS-specific state, and instead the conditions were promoting some adsorbed state that results in the disruption of the homogeneity of the monolayer:protein system, resulting in data that is not suitable for analysis. Reinitializing the system directly at 23 mN/m to avoid expansion resulted in minimal protein insertion, similar to data collected at 25 mN/m.

We then attempted to promote binding by reducing the ionic strength of the buffer with the NaCl concentration decreased from 150 mM to 0 mM (resulting in a 10 mM HEPES only, pH 7.2 buffer) and pre-incubating hTIM3 with Ca^{2+} to reduce protein:Ca^{2+} equilibration time in the trough. Under these conditions, we found stable and reproducible XR binding curves that aligned with the curves of a membrane-inserted state of the murine TIMs (**Fig. 6.14**), as shown by the narrowing and increased intensity of the first peak of the curve.

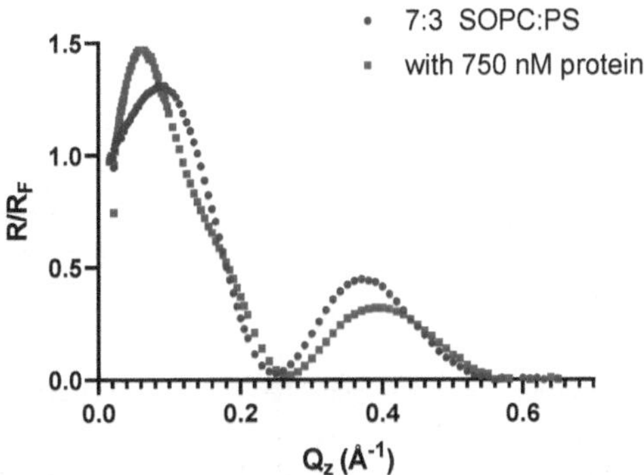

Fig. 6.14. XR reflectivity curve showing binding of hTIM3 to a 7:3 SOPC:PS monolayer at a surface pressure of 23 mN/m. Data were collected at 4 mM Ca^{2+} and in 10 mM HEPES, 0 mM NaCl buffer. Data are plotted as Fresnel Normalized Intensity (*R/R_F*) against the momentum transfer along the z direction (*Q_z*).

One concern is that the low ionic strength would result in non-specific association of the protein with the monolayer. To test this, we performed similar scans on a monolayer composed of 100% SOPC, where we found no evidence of binding or membrane-association (**Fig. 6.15a**). Additionally, to test the Ca^{2+}-specificity of the hTIM3 binding to the 7:3 SOPC:PS monolayer, we added 2 mM EDTA to chelate the Ca^{2+} in the system, resulting in the recovery of the lipid-only curve (**Fig. 6.15b**). Of note, as the addition of EDTA increases the ionic strength of the system (resulting in ionic conditions similar to a buffer with 150 mM NaCl, where we observed minimal binding), we do not necessarily expect its effect to be entirely attributed to the chelation of Ca^{2+} as the ionic strength of the buffer may be contributing to the stripping of the protein from the membrane. However, it is unlikely that its effect is purely due to changes in the ionic strength of the solution, pointing to the Ca^{2+}-specificity of hTIM3 binding.

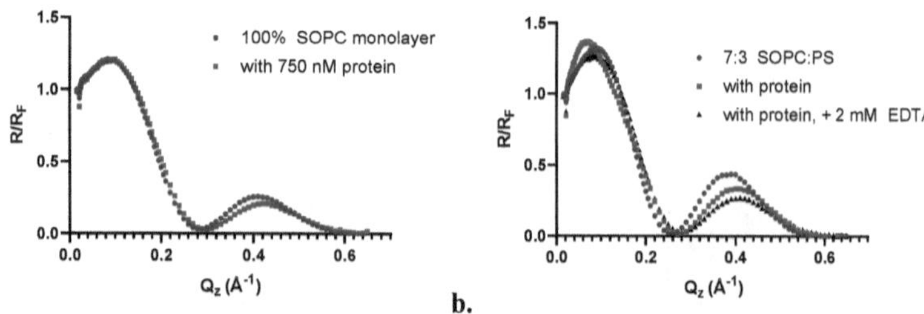

Fig. 6.15. a. XR reflectivity curve showing minimal binding of hTIM3 to a 100% SOPC monolayer at a surface pressure of 23 mN/m. **b.** XR curves showing recovery of initial reflectivity data of the lipid-only monolayer following the addition of 2 mM EDTA to a 7:3 SOPC:PS monolayer with hTIM3 and Ca^{2+} ions introduced. Data were collected at 4 mM Ca^{2+} and in 10 mM HEPES, 0 mM NaCl buffer. Data are plotted as Fresnel Normalized Intensity (***R/R_F***) against the momentum transfer along the z direction (***Q_z***).

While we cannot rule out the possibility that the low ionic strength of the buffer is resulting in nonspecific ionic interactions between the protein and the negatively charged PS, the overall characteristics of the binding data and XR curves (narrowed peak at lower ***Q_z*** ranges, PS-specificity, Ca^{2+} specificity) point to a true membrane-bound hTIM3 state that is suitable for structure analysis. The set of XR curves that were analyzed below were taken in a 10 mM HEPES, pH 7.2 buffer with 2 or 4 mM Ca^{2+}, on 7:3 SOPC:PS monolayers compressed to 23 mN/m, where hTIM3 was pre-incubated with 4 mM Ca^{2+} for 2 min.

Our work with the hTIM3:monolayer system suggests that the protein is relatively unstable. We found that minimal disruptions tolerated by the murine TIM systems would result in rapid aggregation/loss of homogeneous structure in the hTIM3 samples. For example, addition of protein to the subphase of the system or other unavoidable factors, such as dust or bubbles, would often cause runaway expansion. The XR curves found from these samples were consistent with the displacement of lipids by the protein, pointing towards surface activity that is not representative of a membrane-bound state. As we were unable to successfully expand the monolayer to lower surface pressures, we speculate that these results may be due to changes in the spacing within the monolayer that promote adsorption of the protein to the interface that allows for the displacement of lipid.

These experimental features warrant caution and curiosity when studying the membrane-binding of hTIM3. For example, as discussed in **Chapter 6.4**, MD data suggest that hTIM3 may increase local concentrations of PS near the pocket. The clustering of the PS may, in turn, interact differently with free Ca^{2+}, as indicated by the disruption of the membrane at higher calcium concentrations. In this way, hTIM3 may play a functional role in reorganization of the membrane which should be considered within experimental design. Similarly, as hTIM3 is deglycosylated prior to our XR measurements, the aggregation propensity of the protein may point to a functional role of glycosylation sites in stabilizing the solubilized state of the IgV domain. Further work with monolayer and bilayer systems can reveal additional details behind the differences in membrane binding between the murine and human variants of the TIM family.

6.5.2. Resolving hTIM3 Structure at the Lipid Membrane from XR

Our analysis of the XR data aims to address two features of the structural characterization of hTIM3. First, we seek experimental validation for the horizontal orientation of the protein at the membrane found from MD simulations. Second, we want to decipher if the electron density profiles of the various states of the protein (pre and post strand rearrangement) are sufficiently different to observe a difference in the fit of XR data.

Following the methods described in **Chapter 3.3.3.2** and reference 7, we first generated electron density profiles of structures used within the above MD simulations for an array of the Euler angles θ and φ (**Fig. 3.9**), based on PDBs **5F71** and **7M3Y**, along with State A. In choosing **5F71** and

7M3Y, we aimed to have representative structures from both groups of crystal structures for which we have MD simulation data. In comparison, State A was chosen as it had not yet undergone any membrane:protein interactions outside of its pocket engagement with a PS headgroup, and the intent was to see if the XR fit for the most probably orientation of the protein at the membrane corresponds to the final orientation observed from MD simulations displayed by State C. The electron density profiles were then fit to reflectivity data, with the final fits for each orientation plotted as a map of the level of confidence, p-values, computed from the χ^2 distributions for each of the angles. This form of analysis is based on the developed methodologies and the statistical analyses of Daniel Kerr.[7] Within these plots, a p-value less than 0.05 corresponds to an orientation that represents a true state with a 95% confidence.

As the fits of all three structures simultaneously provide information on the orientation and the overall backbone configuration of the protein, we analyzed the fits concurrently for both features. **Fig. 6.16** displays the best fit orientations for each of these three states, with their corresponding maps of the p-values. To aid in visualizing the protein at the membrane, the proteins are rotated through the Euler angles from the best fit and residues that would insert into the membrane are shown in bond form. The protein coverage of the membrane and the depth of the protein's insertions are summarized in **Table 6.4**.

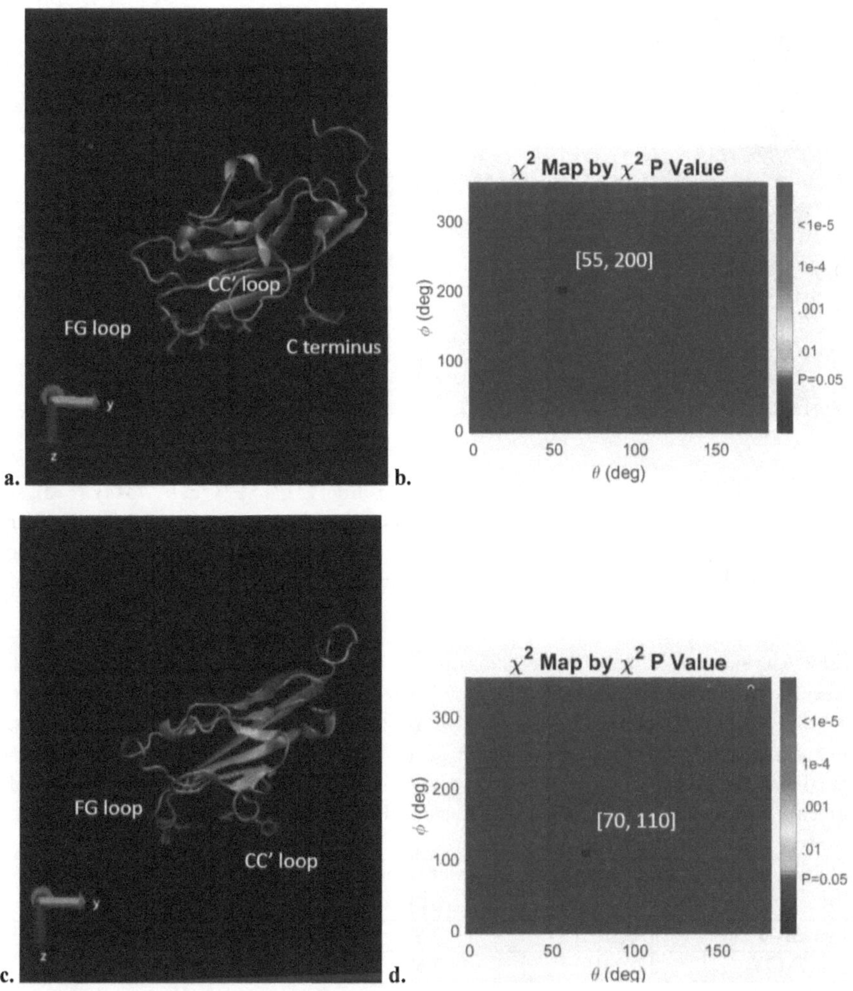

Fig. 6.16. Best fit orientations of structures from PDBs **5F71**.

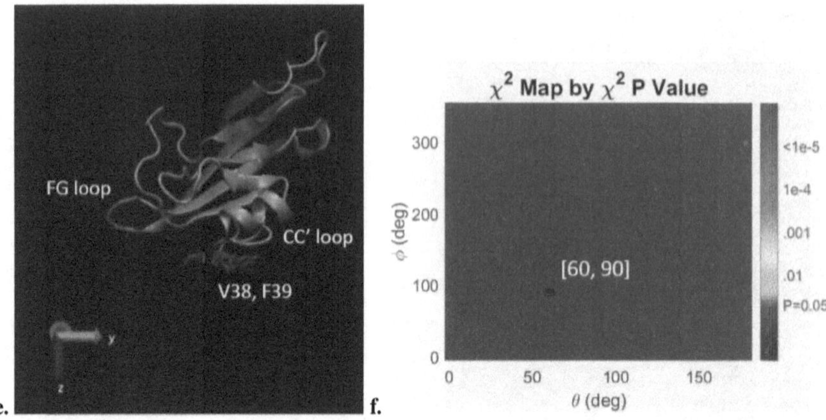

Fig. 6.16, continued. Best fit orientations of structures from PDBs **5F71 (a, b)**, **7M3Y (c, d)**, and State A **(e, f)**. For each pair, the left structure displays the protein rotated through the angles found from the best fit from the map on the right. Within the representation, the lipid membrane direction would be down the z axis with residues that would lie below the lipid headgroups colored in red.

Table 6.4. Summary of best-fit orientations of XR curves found from electron density profiles generated from 5F71, 7M3Y, and State A. Best fit was determined as the orientation with the lowest *p-value* computed from the χ^2 distributions. Depth of insertion was calculated, as described[7], based on the orientation. Determination of whether the orientation was physiologically reasonable was based on if the FG and CC' loops faced the membrane to allow for Ca^{2+} and PS-specificity in the protein's binding. Coverage represents the fraction of the membrane occupied by protein.

Structure	Best Fit [θ,φ]	Depth of Insertion, d_p (Å)	Coverage	Physiologically reasonable?
5F71	[55, 200]	7.48	0.298	Yes
7M3Y	[70, 110]	5.65	0.34	Yes
State A	[60, 90]	3.76	0.265	Yes

As shown in **Fig 6.16**, the fit of each of the three structures yields a single best fit orientation that is a physiologically-reasonable for pocket engagement with PS and Ca^{2+} at the membrane. All three best-fit states place the protein at the membrane with the FG and CC' loops inserted into the membrane, with the exception of the FG loop in State A (**Fig. 6.16e**). However, this can likely be attributed to the open and splayed-out orientation of the FG loop in this state compared to the more closed states in the **5F71** and **7M3Y** structures.

Altogether, the orientations of best fit for the **7M3Y** structure and State A are more closely aligned ($[\theta, \varphi] = [70, 110]$ and $[60, 90]$, respectively), while the fit for **5F71** ($[\theta, \varphi] = [55, 200]$) orients the protein to form contacts between the C-terminus and the membrane (**Fig. 6.16a**). Although we cannot definitively exclude the fit for **5F71** as a possible state despite this contact, we can compare the overall backbone orientation of the fits for **7M3Y** and State A with those of mTIM3 (**Fig. 6.17a**).

Through a similar experimental and fitting process, the best-fit orientation of mTIM3 was found to be $[\theta, \varphi] = [135, 150]$ with a depth of insertion of 12.5 Å[7] (**Fig. 6.17b**). This fit places the protein much deeper into the membrane with near full insertion of the FG and CC' loops. Furthermore, the protein appears to be in a more horizontal orientation than mTIM3 though the orientation difference is not as drastic as those found in State C (**Fig. 6.9**) through the extension of MD simulations from State A. An overlay with **5F71** similarly places most of the protein density along the same axis as the fits for the **7M3Y** structure and State A. Altogether these orientations point towards hTIM3 binding to the membrane in a more horizontally-tilted state in comparison to mTIM3, but the tilt is not as drastic as MD simulations suggest.

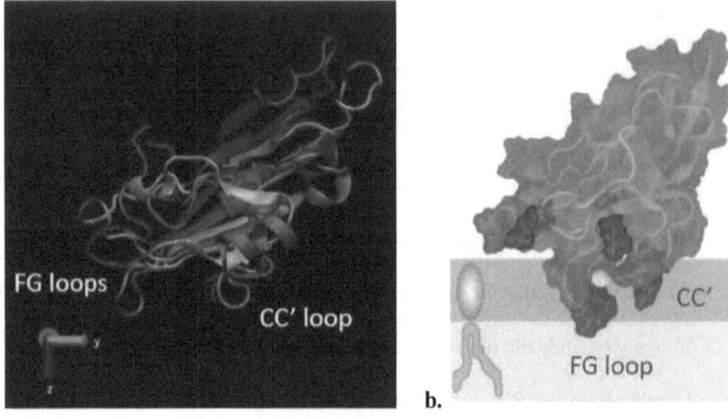

Fig. 6.17. a. Structures of 7M3Y (green) and State A (red) used for fitting rotated corresponding to their best fit orientations (**Table 6.4**). Membrane direction is down the z direction with the membrane sitting in the xy plane. **b.** Representative image of best fit of mTIM3 rotated to its best fit orientation of $[\theta, \varphi] = [135, 150]$ with a depth of insertion of 12.5 Å. Figure adapted from Daniel Kerr.[7]

In our group's work with the mTIM3 system, we have previously found that corroboration of experimental fits with MD simulations occurred only after the published crystal structures were first equilibrated through MD simulations. This was partially due to sidechain relaxation from the crystal-packed state and partially due to the opening of the FG-CC' cleft. In our fits of electron density profiles based on the crystal structure of **7M3Y**, we were concerned that similar differences between equilibrated and non-equilibrated states would manifest as an alternate fit to the reflectivity data. To address this, we used representative frames from Trials 4 and 5 of the membrane-docked simulations of **7M3Y** from the horizontal orientation (**Chapter 6.4.3.1**), where we intentionally chose one of the trials that had a near identical backbone alignment with the initial structure (Trial 4) and the trial that was beginning to undergo strand rearrangement (Trial 5, **Fig. 6.11**). The resulting best fit orientations and corresponding depths of insertion are shown in **Table 6.5** and **Fig. 6.18**.

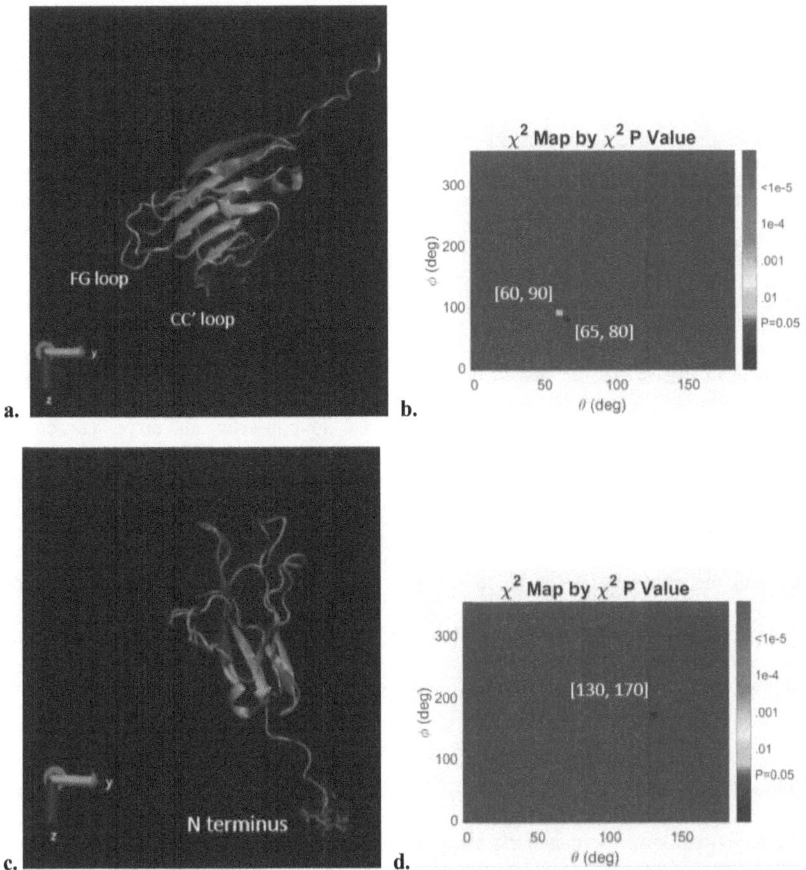

Fig. 6.18. Best fit orientations of representative frames from Trial 4 (**a, b**) and Trial 5 (**c, d**) of membrane-docked simulations of **7M3Y** in the horizontal orientation. For each pair, the left structure displays the protein rotated through the angles found from the best fit from the map on the right. Within the representation, the lipid membrane direction would be down the z axis with residues that would lie below the lipid headgroups colored in red.

143

Table 6.5. Summary of best fit orientations of XR curves found from electron density profiles generated from 5F71, 7M3Y, and State A. Best fit was determined as the orientation with the lowest *p-value* computed from the χ^2 distributions. Depth of insertion was calculated, as described[7], based on the orientation. Determination of whether the orientation was physiologically reasonable was based on if the FG and CC' loops faced the membrane to allow for Ca^{2+} and PS-specificity in the protein's binding. Coverage represents the fraction of the membrane occupied by protein.

7M3Y Trial	Best Fit $[\theta,\varphi]$	Depth of Insertion, d_p (Å)	Coverage	Physiologically reasonable?
Trial 4	[65, 80]	3.42	0.30	~
Trial 5	[130, 170]	8.01	0.32	No

As evident in **Fig. 6.18a**, the best fit orientation of Trial 4 barely places the protein below the surface of the membrane with only a portion of the CC' loop inserting and no residues on the FG loop. While this state is even more horizontally tilted, the lack of insertion of residue F39 suggests that this orientation may not be physiologically relevant as across our MD of hTIM3 and previous work with the murine TIMs,[13] the insertion of bulky, hydrophobic residues plays a role in stabilizing the protein at the membrane. However, it is interesting to note that the map of best fits for this structure yielded a secondary minimum at $[\theta, \varphi] = [60, 90]$, the exact orientation found for the best fit of State C.

In comparison, the best fit orientation of Trial 5 yields a non-physiologically relevant orientation, where the N-terminus inserts into the membrane. Not only would this bound state be non-PS or Ca^{2+}-specific, within a physiological system the mucin stalk at the N-terminus would prohibit this orientation. It is likely that the electron density from the unfolding of strands C' and C", seen in **Fig. 6.11**, is contributing to fit of the protein, indicating that the current state of Trial 5 is not a true representative of the bound-state for the protein. It is possible that future equilibration of this trial will result in a protein structure that is amenable to XR fitting. While we do not gain any insight into the fitting from Trial 5, the secondary minimum of $[\theta, \varphi] = [60, 90]$ found in the fitting of

Trial 4 aligning with the best fit orientation found for State A is promising for the future collection and fitting of XR data to converge on this, or a similar, orientation.

Similarly, we can use states from the HMMM simulations of State A to observe if the additional membrane contacts alter the best fit orientation. For this, we fit States B and C to the XR data as representatives of the changes that the protein undergoes over 200 ns or 1 µs of simulation time. As shown in **Fig. 6.19** and **Table 6.6**, the found orientation of State B (**Fig. 6.19a,b**) , $[\theta, \varphi] = [60, 100]$ aligns with the found orientation from State A, $[\theta, \varphi] = [60, 90]$, placing the CC' loop more deep into the membrane in comparison to State A. Of note, the values for depth of insertion (3.8 Å vs. 8.6 Å for State A and B, respectively) cannot be directly compared between the two, as residue F39 lies flat against the protein in State A but extends away from the protein in State B, where it can form hydrophobic contacts with lipids in the membrane. While we cannot identify a single best fit orientation between States A and B, the alignment of inserted residues (V38, F39, and E40) between the two states that reflects these contacts is promising that State B is reinforcing the orientations found through the fits of State A.

In comparison, the fit of State C yields an orientation of $[\theta, \varphi] = [15, 330]$ and places the FG and BC loops deep into the membrane (**Fig. 6.19c,d**). Though the FG and CC' loops still face membrane, the hydrophobic residues of the CC' loop do not insert into the membrane, and this orientation likely lacks Ca^{2+} and PS specificity, both of which indicate that this fit does not represent a true membrane-bound state. Altogether, the fit results of State C suggest that the observed horizontal tilt and the rotation of the FG loop along with other minor changes to the protein's backbone structure across 1 µs of simulation time (**Fig. 6.9b**) are not reflected in the XR data and may represent a limitation of the HMMM system for longer simulations of the hTIM3:membrane system.

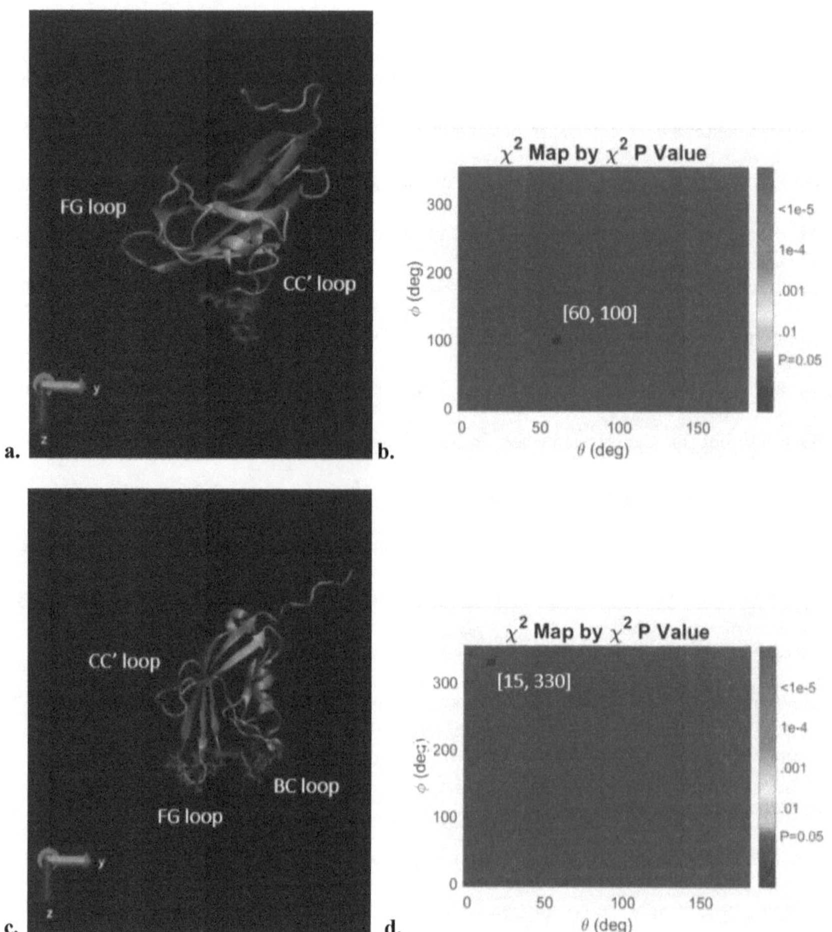

Fig. 6.19. Best fit orientations of representative frames from HMMM simulations of State A after 200 ns (**a, b**) and 1 µs (**c, d**). For each pair, the left structure displays the protein rotated through the angles found from the best fit from the map on the right. Within the representation, the lipid membrane direction would be down the z axis with residues that would lie below the lipid headgroups colored in red.

146

Table 6.6. Summary of best fit orientations of XR curves found from electron density profiles generated from State B and State C. Best fit was determined as the orientation with the lowest *p-value* computed from the χ^2 distributions. Depth of insertion was calculated, as described[7], based on the orientation. Determination of whether the orientation was physiologically reasonable was based on if the FG and CC' loops faced the membrane to allow for Ca^{2+} and PS-specificity in the protein's binding. Coverage represents the fraction of the membrane occupied by protein.

State, extension time	Best Fit [θ,φ]	Depth of Insertion, d_p (Å)	Coverage	Physiologically reasonable?
State B, 200 ns	[60, 100]	8.6	0.25	Yes
State C, 1 μs	[15, 330]	10.2	0.31	~

An additional concern of ours was that electron density contributions from the N-terminal tail were artificially placing the protein in a non-representative state, resulting in the poor fits of structures such as State C or Trial 5 of **7M3Y**. The N-terminal region of our protein construct likely does not form secondary structure, as we did not observe its formation across ~20 μs simulation time nor through the AlphaFold prediction tool and is the part of the protein that connects to the mucin stalk of the rest of the protein. However, as shown in **Fig. 6.9c**, the N-terminus is flexible across the simulations despite only minor changes to overall protein backbone structure, and so choosing a single representative state of the tail is likely underrepresenting the smearing of the measured electron density that occurs from the fluctuations of the protein in our experimental protein:monolayer system.

To address this concern, we generated electron density profiles of State A but replaced the more collapsed tail with the extended tail from State C. We then fit the same XR data sets using the new electron density profiles. From this we found that the best fit orientation was still [θ, φ] = [60, 90] with only a minor change to depth of insertion (d_p = 3.54 Å vs. 3.76 Å). A similar analysis was performed for the structures generated from **5F71** and **7M3Y**, which also showed little change in fitting. Altogether, the minor differences between the two tails suggest that the electron density contributions from the N-terminal tail minimally affect the overall best fit orientation.

Altogether, the electron density profiles of hTIM3 place the protein in a more horizontal orientation in comparison to mTIM3, as evident through the fits of State A and **7M3Y**. Though not as drastically horizontal as observed by MD, this tilted orientation may have critical implications for hTIM3's ability to undergo strand rearrangement along with other physiologically-relevant protein functions, such as protein:protein contacts.

However, our preliminary fits of the various protein structures are unable to resolve differences in the protein's backbone arrangement of the C' and C" strands, as evident by similar fits for State A and structures generated from PDB **7M3Y**. While this may indicate a limitation within the fitting method's sensitivity to differences in electron density differences between states, it is also possible that these differences may be resolved through the additional collection of reflectivity data of the protein:monolayer system as previous work by our group has found that additional analytical methods, such as bootstrapping, can aid in the convergence of data to a single orientation given enough XR curves.[7] We are currently unable to use these methods due to limited sampled data, but our results are promising for future work to uncover further detail of the hTIM3-membrane bound state.

6.6. Conclusions and Future Directions

Our work here demonstrates the importance of considering protein:lipid contacts in the hTIM3:membrane system. We show that these contacts contribute to a horizontally bound orientation of the protein that differentiates the protein from its murine counterpart, partially related to R47:PS contacts and the splaying out of the FG loop. Through structural comparison of the various crystal structures and corresponding MD simulations, we additionally identify key features and states of the protein that may participate in a global conformational switch when bound to a lipid-membrane, including the formation of the Y55-W61 contact and the breaking of E40-R47-Q91 triad. While we have yet to isolate a representative structure of hTIM3 at the membrane from our preliminary results, we highlight the importance of studying the protein:membrane system through a multi-pronged approach.

Based on the data presented in this chapter, future work within this project will aim to extend the current membrane-docked MD simulations of hTIM3 to µs timescales and the initialization of a third set of trials for representation of the crystal structures in "Group 1." The primary goal of this work would be to capture the full conformational change on the membrane and to connect the various proposed solution state structures to the protein's membrane-bound state. Along with MD simulations, additional collection of XR data can further refine the best fit orientation of hTIM3 in its monolayer-bound state.

The broader aim of this work was to contribute to the progress within the field of therapeutic development in the drug targeting of the IgV domain of hTIM3. Recently, Ma et al.[14] have identified a small molecule, ML-T7, that binds to the FG-CC' cleft and has functional anti-tumor activity. In comparison, Rietz at al.[11] have identified several small molecules that bind to C" strand that would likely interfere with the strand rearrangements discussed above. Our work here suggests that there is room within drug development to optimize the state-selectivity of these small molecules. For example, one could selectively target a PS-bound state to retain hTIM3's PS-binding functionality while preventing conformational changes that may be linked to downstream responses. Similarly, dual targeting of both the FG-CC' cleft and the C" strand could have a greater therapeutic effect on the anti-tumor activity and survival of T cells. Altogether, our work highlights the importance of considering how the PS-binding role of hTIM3 alters its structural state.

6.7. References

1. Kuchroo, V. K., Umetsu, D. T., DeKruyff, R. H. & Freeman, G. J. The TIM gene family: emerging roles in immunity and disease. *Nat Rev Immunol* **3**, 454–462 (2003).

2. Weber, J. K. & Zhou, R. Phosphatidylserine-Induced Conformational Modulation of Immune Cell Exhaustion-Associated Receptor TIM3. *Sci Rep* **7**, 13579 (2017).

3. Smith, C. M., Li, A., Krishnamurthy, N. & Lemmon, M. A. Phosphatidylserine binding directly regulates TIM-3 function. *Biochemical Journal* **478**, 3331–3349 (2021).

4. Kerr, D. *et al.* How Tim proteins differentially exploit membrane features to attain robust target sensitivity. *Biophys J* **120**, 4891–4902 (2021).

5. Han, G., Chen, G., Shen, B. & Li, Y. Tim-3: An Activation Marker and Activation Limiter of Innate Immune Cells. *Front Immunol* **4**, (2013).

6. Gandhi, A. K. *et al.* High resolution X-ray and NMR structural study of human T-cell immunoglobulin and mucin domain containing protein-3. *Sci Rep* **8**, 17512 (2018).

7. Kerr, D. The Membrane Context of Phosphatidylserine Exposure Influences its Recognition by the TIM Proteins: A Structurally Guided Study. (University of Chicago, 2019).

8. LIU, N. *et al.* Crystal structures of human TIM members: Ebolavirus entry-enhancing receptors. *Chinese Science Bulletin* **60**, 3438–3453 (2015).

9. Zhang, D. *et al.* Identification and characterization of M6903, an antagonistic anti–TIM-3 monoclonal antibody. *Oncoimmunology* **9**, (2020).

10. Pagliano, O. *et al.* Tim-3 mediates T cell trogocytosis to limit antitumor immunity. *Journal of Clinical Investigation* **132**, (2022).

11. Rietz, T. A. *et al.* Fragment-Based Discovery of Small Molecules Bound to T-Cell Immunoglobulin and Mucin Domain-Containing Molecule 3 (TIM-3). *J Med Chem* **64**, 14757–14772 (2021).

12. Hokanson, D. E. & Ostap, E. M. Myo1c binds tightly and specifically to phosphatidylinositol 4,5-bisphosphate and inositol 1,4,5-trisphosphate. *Proceedings of the National Academy of Sciences* **103**, 3118–3123 (2006).

13. Kerr, D. *et al.* Sensitivity of peripheral membrane proteins to the membrane context: A case study of phosphatidylserine and the TIM proteins. *Biochimica et Biophysica Acta (BBA) - Biomembranes* **1860**, 2126–2133 (2018).

14. Ma, S. *et al.* Identification of a small-molecule Tim-3 inhibitor to potentiate T cell–mediated antitumor immunotherapy in preclinical mouse models. *Sci Transl Med* **15**, (2023).